Praise for

MORE EVERYTHING FOREVER

"This is a really important contribution to our discussion of the future and what it might hold, and what we should be trying for now. Some of these popular ideas about the future are foolish enough to distort our current reality, and they deserve to be revealed as such. Adam Becker's book is very entertaining as it exposes how the emperor has no clothes. . . . This is an important book as well as a good one."

—Kim Stanley Robinson, author of *The Ministry for the Future*

"With a wild and utterly engaging narrative, Becker gives us a refreshing reality check on the fantasies of billionaires, futurists, and utilitarian philosophers who are plotting to 'optimize' the future of humanity. A fascinating exposé of the extreme techno-solutionism promoted by the most powerful and influential technocrats of our era."

—Melanie Mitchell, computer scientist and author of *Artificial Intelligence*

"*More Everything Forever* dismantles the toxic techno-optimism endemic in Silicon Valley and outlines why the most pressing problems society faces can't be solved with technology alone. The book is a must-read for understanding why the visions of the future promoted by today's techno-oligarchs are built on pseudoscience and far-fetched fantasies mixed with racism, eugenics, and colonialism. Becker argues that focusing on invented future problems that may or may not ever come to pass gives techno-optimists license to neglect urgent and real problems like global warming and income inequality that are threatening humanity in the here and now. *More Everything Forever* feels particularly urgent and timely as billionaires like Elon Musk and Jeff Bezos vie for political power."

—Christie Aschwanden, author of *Good to Go*

"I love this book. We are sitting astride an inflection point. The world as we know it or think we know it is rapidly vanishing, replaced by something. . . . Well, we're really not sure what. It used to be that we could point to an Orwellian future where 2 + 2 = 5 and so on and so forth. Perhaps disturbing but still understandable. But nowhere in this can of worms is there a vision of the subjugation of everything to a dystopia of machines. As Becker reports, we will all bow down before the specter of advanced technology before the onrushing oligarchy. The only remaining question: Will it be an oligarchy of advanced devices or just nasty, self-serving, delusional billionaires equipped with advanced iPhones? Becker has become our greatest prophet of doom."

—Errol Morris, documentary filmmaker and author of *The Ashtray*

"Our world has fallen into the clutches of billionaires who mistake dystopian science fiction stories for suggestions, rather than warnings. Speaking in my capacity as a dystopian science fiction writer, I can confirm that this isn't merely very *stupid*, it's also *very, very bad.*"

—Cory Doctorow, author of *Red Team Blues*

MORE
EVERYTHING
FOREVER

MORE EVERYTHING FOREVER

AI OVERLORDS, SPACE EMPIRES, AND SILICON VALLEY'S CRUSADE TO CONTROL THE FATE OF HUMANITY

ADAM BECKER

BASIC BOOKS
New York

Basic Books
Hachette Book Group
1290 Avenue of the Americas, New York, NY 10104
www.basicbooks.com

Printed in Canada

First Edition: April 2025

Published by Basic Books, an imprint of Hachette Book Group, Inc. The Basic Books name and logo is a registered trademark of the Hachette Book Group.

The Hachette Speakers Bureau provides a wide range of authors for speaking events. To find out more, go to hachettespeakersbureau.com or email HachetteSpeakers@hbgusa.com.

Basic books may be purchased in bulk for business, educational, or promotional use. For more information, please contact your local bookseller or the Hachette Book Group Special Markets Department at special.markets@hbgusa.com.

The publisher is not responsible for websites (or their content) that are not owned by the publisher.

Print book interior design by Jeff Stiefel.

Library of Congress Cataloging-in-Publication Data

Names: Becker, Adam, author.
Title: More everything forever : AI overlords, space empires, and Silicon Valley's crusade to control the fate of humanity / Adam Becker.
Description: New York : Basic Books, 2025. | Includes bibliographical references and index.
Identifiers: LCCN 2024031543 (print) | LCCN 2024031544 (ebook) | ISBN 9781541619593 (hardcover) | ISBN 9781541619609 (ebook)
Subjects: LCSH: Technological innovations. | Artificial intelligence.
Classification: LCC HD45 .B393 2025 (print) | LCC HD45 (ebook) | DDC 658.5/14—dc23/eng/20250118
LC record available at https://lccn.loc.gov/2024031543
LC ebook record available at https://lccn.loc.gov/2024031544

ISBNs: 9781541619593 (hardcover), 9781541619609 (ebook)

MRQ-T

1 2025

For my parents,
who taught me to question everyone,
even them.

The author acknowledges with gratitude
the support of the Alfred P. Sloan Foundation
in the research and writing of this book.

There is a reason, after all, that some people wish to colonize the moon, and others dance before it as before an ancient friend.

—James Baldwin, *No Name in the Street*

How you play is what you win.

—Ursula K. Le Guin, "The Matter of Seggri"

Sci-Fi Author: In my book I invented the Torment Nexus as a cautionary tale

Tech Company: At long last, we have created the Torment Nexus from classic sci-fi novel Don't Create The Torment Nexus

—Alex Blechman, on Twitter

CONTENTS

INTRODUCTION

The dream is always the same: go to space and live forever. "Perfect health, immortality, yada yada yada," Eliezer Yudkowsky tells me. "Transhumanism, transcension, yada yada yada yada. That's just the obvious stuff. Just include the glorious transhumanist future. . . . Maybe we can do better than whatever scenario but at least that's the minimum."[1]

Yudkowsky is the cofounder of the Machine Intelligence Research Institute (MIRI), a controversial artificial intelligence think tank. He's telling me about the future he'd like to see if an AI much smarter and more capable than a person can be built and made to serve us, keeping its superhuman abilities under human control. Such a machine could bring us an entirely different way of life. "If you imagine something that's worse than mansions with robotic servants for everyone, you are not being ambitious enough," he says. But the "glorious transhumanist future" that he is alluding to goes well beyond that. Transhumanism is the belief that we can and should use advanced technology to transform ourselves, transcending humanity and becoming something more. That generally involves finding ways around the limits of the human body—ending illness, aging, and death—as well as increasing intelligence and other mental capacities. But it also carries with it various promises about the future of humanity (or transhumanity) itself: that our fate lies in the stars, that we will build an intergalactic civilization, that we will reshape the universe to our desires just as we will have reshaped ourselves.

Yudkowsky thinks this future is desirable but not inevitable. What's inevitable, he says, is the advent of artificial general intelligence (AGI): machines that can outperform humans at any task. AGI is "something that is sufficiently better at predicting and steering the future that it can beat humanity on the grounds we've always claimed as our own," Yudkowsky tells me. He thinks such a machine may not be far off. "My sense is that we are zero to two breakthroughs" away from AGI, he says. He refuses to speculate in detail on when an AGI might be built—"How would I know that, man? I ain't no oracle!"—but he is confident that it will arrive, perhaps in "more like five years than fifty years."[2] And when it does, he says, it will set about making itself more powerful, gaining control of more resources and more computing power to increase its own intelligence. The problem that Yudkowsky sees is that, while AGI is inevitable, its servitude is not. An AGI's goals, he fears, won't be the same as ours. Whatever those goals are, says Yudkowsky, humanity will almost certainly be in its way—so it'll kill us all once it has found a surefire way to do so.

The idea of AGI is taken quite seriously by many people in the tech industry, as are Yudkowsky's concerns and desires. Ever since a new generation of AI caught the public imagination with the launch of ChatGPT in late 2022, some AI researchers and tech executives have been warning journalists and government officials about the "existential threat" that out-of-control AGI could pose to humanity. Without careful regulation and industry agreements, they say, Yudkowsky's worst fears could be realized. Yudkowsky himself goes further: He wants all advanced AI research shut down immediately, via international agreement, until there is a method to ensure any future AGI is unlikely to wipe out humanity. And he wants that shutdown enforced with a nuclear threat. "Make it explicit in international diplomacy that preventing AI extinction scenarios is considered a priority above preventing a full nuclear exchange, and that allied nuclear countries are willing to run some risk of nuclear exchange," he wrote in *Time* magazine in 2023.[3]

Not everyone in AI agrees with Yudkowsky about these purported dangers, and fewer agree that a total shutdown of AI research is a good idea. It's unlikely that Yudkowsky could broker an international agreement anyhow—he doesn't have enough political influence to do that. But his ideas do carry weight with some of the politically connected leaders of the tech industry. One of them is Sam Altman, the CEO of OpenAI, the company behind ChatGPT; he's suggested that Yudkowsky may eventually "deserve the Nobel Peace Prize" for his work on AI.[4]

Altman doesn't want to shut down AI research, but he agrees with Yudkowsky that AGI is inevitable and could be coming soon. He also agrees that once AGI does arrive, it will be able to improve itself, leading quickly to a radically different future. "The technological progress we make in the next 100 years will be far larger than all we've made since we first controlled fire and invented the wheel," Altman claims. "This technological revolution is unstoppable. And a recursive loop of innovation, as these smart machines themselves help us make smarter machines, will accelerate the revolution's pace." In an essay on his personal website, Altman declares that this inexorably approaching future will involve computer programs doing "almost everything, including making new scientific discoveries that will expand our concept of 'everything.'" Altman envisions AI taking over all services, manufacturing, and production. "Imagine a world where, for decades, everything—housing, education, food, clothing, etc.—became half as expensive every two years," he writes. "This revolution will create phenomenal wealth . . . [but] the world will change so rapidly and drastically that an equally drastic change in policy will be needed to distribute this wealth and enable more people to pursue the life they want."

In his essay, Altman lays out a vision of "capitalism for everyone" as the solution to the policy problem he describes. "The best way to improve capitalism is to enable everyone to benefit from it directly as an equity owner," he writes. To accomplish this, he proposes that the very wealthiest companies give a small amount of their value to the public each year. This would be accomplished through a tax on those companies, paid in shares, which would then be distributed evenly among the American public. That tax must be much smaller than the average growth rate of the companies—but, Altman assures us, "once AI starts to arrive, growth will be extremely rapid," allowing the tax rate to be high enough to provide substantial income for Americans through shares of the companies so taxed. "Poverty would be greatly reduced and many more people would have a shot at the life they want," Altman writes. "If everyone owns a slice of American value creation, everyone will want America to do better. . . . The new social contract will be a floor for everyone in exchange for a ceiling for no one, and a shared belief that technology can and must deliver a virtuous circle of societal wealth." As a parenthetical aside, Altman notes that "strong" government will still be needed "to make sure that the desire for stock prices to go up remains balanced with protecting the environment, human rights, etc." He goes on to consider details of how this plan might be implemented—he suggests making the tax part of the US Constitution—and argues that a "pro-business and pro-people" plan like his would be politically popular. "The changes coming are unstoppable," Altman's essay concludes. "If we embrace them and plan for them, we can use them to create a much fairer, happier, and more prosperous society. The future can be almost unimaginably great."[5]

Altman's policy proposals seem outlandish, but they carry some weight, if only because of his access to government. Altman has testified before the US Senate and met with Joe Biden while he was president. A few months after the explosive launch of ChatGPT, Altman went on a world tour, meeting with political leaders and venture capitalists in dozens of countries to discuss the present and future of AI. Around that same time, the *New York Times* wrote that Altman's "grand idea" was that his company, OpenAI, "will capture much of the world's wealth through the creation of AGI and then redistribute this wealth to the people."[6] With that lens, his essay takes on a new meaning. Altman apparently wants to make the United States into one enormous company town, with shares in OpenAI replacing the dollar. The US government would become, in effect if not in law, a division of the company, responsible for disbursing company dollars to us, the public. This would, Altman hopes, encourage us to think of OpenAI's success as America's success and as our own—Altman explicitly makes this identification in his essay. All products would come from OpenAI in his proposed future, because in that future AI does literally everything, meaning that the company dollars can only be spent at the company store. (Those company dollars are ostensibly shares, but since the amount given out each year is capped below the growth rate of the company, Altman and the board would always retain control of OpenAI, and the shares owned by the American public would never get anywhere near an appreciable fraction of company ownership.) Thus, Altman's promise of goods halving in price every two years would depend solely on his goodwill, because things will cost whatever Altman and the OpenAI board want them to cost. This is a proposal for total capture of the national economy, making Altman functionally the king of the United States and possibly the world. He has been quite explicit about replacing money: in 2024, he said that in the future, instead of a universal basic income, there might be "universal basic compute," allocating people time on a future GPT model that can (somehow) produce anything they need.[7] The page on Altman's website hosting his essay contains a surprisingly clear indication of this dream. At the top of the page, there's an interactive illustration of a pile of dollar bills. Hover your mouse over them, and the dollars turn into computer chips.

* * *

Altman's power fantasies and Yudkowsky's nightmares are pieces of a bigger picture of the future, one shared by many of the wealthiest and most influential people in the tech industry. That future is straight out of science fiction: people's minds uploaded into computers to live for all eternity in a silicon paradise, watched over by a benevolent godlike AI; a ceaselessly expanding empire spanning the stars, disassembling planets, and consuming galaxies; all needs satisfied, all fears assuaged, all desires sated through the power of unimaginably advanced technology.

The tech billionaires aren't coy about this. Like Altman, they talk about such futures as inevitable, or the only good option aside from extinction. Jeff Bezos has repeatedly said that he wants a trillion people living in space to enable a future of perpetual growth, lest we "stagnate" here on Earth. Elon Musk has been tweeting for years about the importance of going to Mars and beyond to save humanity. "The true battle is: Extinctionists who want a holocaust for all of humanity, versus Expansionists who want to reach the stars and Understand the Universe," he wrote.[8] And Marc Andreessen wants an eternally triumphal "techno-capital machine" to conquer the cosmos with AI and the power of entrepreneurship.[9] Other tech billionaires have provided millions of dollars to the effective altruism community, which is doing academic work to provide a moral argument in favor of this kind of future. They've also given comparable sums to the rationalist movement, a community that developed around Yudkowsky's fears of an AI apocalypse derailing the glorious future promised by technology.

That future, Altman and his fellow billionaires claim, will be good for everyone. They also claim it's based on sound science, that this is just the future as revealed by a close study of technology and its development. These claims are, at best, deeply questionable. But the tech billionaires and the groups they fund seem to earnestly believe them, despite the evidence against such ideas. That's not a huge surprise—these futures are deeply seductive. They offer the promise of something that even billions of dollars can't buy: transcendence of all limits, even of mortality itself. And in the meantime, before the promised future arrives, its pursuit offers absolution. The credence that tech billionaires give to these specific science-fictional futures validates their pursuit of *more*—to portray the growth of their businesses as a moral imperative, to reduce the complex problems of the world to simple questions of technology, to justify nearly any action they might want to take—all in the name of saving humanity from a threat that doesn't exist, aiming at a utopia that will never come. The carbon footprint of Amazon's shipping network or SpaceX's rocket fleet can't possibly matter as much as hastening the glorious immortal future of humanity in space. And if that future never comes, that just means the excuse of its pursuit will never wear out. If the apocalypse actually arrived, the doomsday cult leader would lose their followers.

More than anything, these visions of the future promise control by the billionaires over the rest of us, just as in Altman's essay. But that control isn't limited to the future—it's here, now. Their visions of the future are news; they inform the limits of public imagination and political debate. Setting the terms of such conversations about the future carries power in the present. If we don't want tech billionaires setting those terms, we need to understand their ideas about the future: their curious origins, their horrifying consequences, and their panoply of ethical gaps and scientific flaws.

1

NOT FADE AWAY

A few years ago, I was sitting at a table outside the undergraduate library on the Berkeley campus on a gorgeous afternoon in late summer, quietly working on a project with a few colleagues who were there with me for a small workshop. Taking a moment to sit and think about what I was doing, my attention—a flighty thing even at the best of times—was caught by a snippet of conversation from the next table over between two students, one older, the other in their first year. The former was trying to sell the latter on a student group promoting effective altruism, a new approach to charitable giving. The older student described it as a fun way to socialize with other students while doing something worthwhile, and he capped off his pitch with a personalized addition for the younger student, who was studying engineering. "A lot of effective altruists are engineers," he said, "because they like to optimize the shit out of things."

I chuckled quietly, and then had to explain to the others at my table why, and what I knew about effective altruism. What I didn't know was that several years earlier, in 2012, a similar conversation had happened on the other side of the country, between Will, a young philosopher specializing in ethics, and Sam, a junior at MIT. Sam was trying to figure out what to do with his life, and at an Au Bon Pain next to Harvard Square, Will pitched him on a central concept of effective altruism: "Earn to give," the idea, roughly, that one of the best ways to make the world a better place is to make a large amount of money, and then donate much of that money to worthy causes that help people.[1] Sam nodded, saying simply, "Yep. That makes sense." He took Will's advice—and his philosophy—and ran with it. He adopted effective altruism wholeheartedly, taking it with him to a job at Jane Street, a Wall Street firm specializing in high-frequency trading. After a few years working there as a trader (he claimed that he donated about half of his salary there to charity), Sam left to take a job alongside Will, as director of business development for the Centre for Effective Altruism (CEA).[2] Shortly after that, Sam's career really took off.

Sadly, today Will and Sam aren't as friendly as they used to be. "I don't know which emotion is stronger: my utter rage at Sam (and others?) for causing such harm to so many people," Will wrote in November 2022, "or my sadness and self-hatred for falling for this deception."[3]

* * *

It's not possible to give a good explanation of what happened between Will and Sam—and what Sam did that angered Will (and many, many others)—without first explaining effective altruism. Effective altruism seems relatively straightforward on the face of it: evaluate the best ways to make the world a better place, and then devote as much money and time as you can to those efforts. The core of the idea goes back to the philosopher Peter Singer, a professor of ethics at Princeton. Singer has advocated that everyone should give all (or very nearly all) of their disposable income to charity.[4] His most famous argument for this idea starts with a simple thought experiment.

Suppose you're taking a walk down a reasonably busy path in a small park. It's a bit chilly outside, so you're wearing a sweater—maybe a pretty nice sweater, cashmere or something like that—and a comfortable yet fashionable coat over it. As you stroll down the road, you come upon a muddy pond, and you immediately see that a child is drowning. In fact, the child appears to be on the verge of sinking entirely into the opaque, slimy waters, barely able to call for help. Nobody else has spotted the child, or if they have, they don't seem to care, hardly slowing down as they walk past. Every second counts, but luckily the water is pretty shallow, so if you hurry, you know you can wade into the pond and save the kid. But you also know that jumping in immediately would probably ruin your nice sweater, your favorite coat, and the rest of your clothes. And it seems like you won't be able to see the child once they slip below the surface of the pond, so your best chance to reach them in time is to jump in immediately without stopping to remove your clothes. What do you do? Do you rescue the child? Is the time and money you'll have to spend replacing your clothes worth saving a life? Or are the clothes and money more important?

Nearly everyone would agree that saving the child is always the right move here and is definitely worth spoiling your clothes. But, Singer argues, if that's so—if a set of clothes is always less valuable than the life of a child—then surely it doesn't matter where the child is. And, as Singer goes on to point out, it is in fact at least as easy for a moderately affluent person to save the life of a real impoverished child in another part of the world as it is to save the life of the hypothetical nearby child drowning in the muddy pond. "We are all in that situation of the person passing the shallow pond," wrote Singer in his 1997 essay "The Drowning Child and the Expanding Circle." "We can all save lives of people, both children and adults, who would otherwise die, and we can do so at a very small cost to us: the cost of a new CD, a shirt or a night out at a restaurant or concert, can mean the difference between life and death to more than one person somewhere in the world—and overseas aid agencies like Oxfam overcome the problem of acting at a distance."[5]

As it turns out, Singer was probably incorrect about the cost of saving one life for these relief agencies. While advertising materials for such charities sometimes state that lives can be saved for less than $100, a more realistic estimate is likely to be around $5,000.[6] But Singer's point still stands, and his thought experiment is easy to modify to account for this: If we say that you happen to be borrowing an exceptionally nice designer jacket from a wealthy friend while you're out on your walk, you'd probably still elect to save the child even if it means replacing your friend's jacket. Or we could imagine a situation where you're out with a large group of friends, and all of you need to jump into a muddy pond to pull a child free, each of you losing several hundred dollars' worth of clothing, cell phones, and other personal items in the process. There are counterarguments to Singer's position, but most of them are aimed at the most extreme conclusion that can be taken, namely that *all* of one's disposable income should be spent on efforts to save lives around the world.[7] Even if that's not true, Singer's argument does suggest that we should be spending more on saving lives than most of us usually do.

While Singer's argument is compelling, his overall approach to ethics is less persuasive. Singer is a utilitarian. Utilitarians believe, roughly, that acting ethically means making choices that lead to the greatest good and least suffering, maximizing an abstract quantity known as "utility." Singer includes animals in this moral calculus. He has advocated against factory farming and in favor of vegetarianism, and his writings on those subjects are quite influential among animal-rights activists. But taking a utilitarian view of ethics, in its most basic form, reduces ethical questions to an optimization problem: What actions will lead to the largest amount of utility in the world? The obvious follow-up questions are about what constitutes "utility" and how to know what actions will promote it, neither of which are clear. Modern utilitarians have introduced nuance into their positions to handle such questions. Nonetheless, sometimes even more sophisticated forms of utilitarianism can lead to troubling—or even abhorrent—conclusions regarding the best way to promote the greater good and avoid suffering. For example, Singer has said that euthanasia of infants with significant disabilities, as well as adults with advanced dementia, is morally acceptable under certain circumstances.

Unsurprisingly, Singer is infamous for his views on euthanasia. Aside from that, he's probably best known for his views on animal rights. But it was his argument for giving more to charity that had the most influence on William MacAskill—Will—and a small group of his fellow philosophers. In 2009, while in graduate school at Oxford, MacAskill cofounded the nonprofit Giving What We Can, along with Toby Ord, another philosopher there. The organization asks members to pledge 10 percent of their income to charity until they retire. MacAskill and Ord signed the pledge themselves. "I was on board with the idea of binding my future self—I had a lot of youthful energy, and I was worried I'd become more conservative over time," recalled MacAskill.[8]

Ord and MacAskill weren't merely interested in donating as much money as they could—they wanted to figure out the best way to donate that money to help people. Ord had been donating 25 percent of his income to charity and had determined the best place to send that money was a foundation treating blindness in developing countries. MacAskill persuaded him that charities aimed at eliminating intestinal parasites were a better choice, pointing to economic research that suggested such charities were a hundred times more cost-effective. Further research suggested that charities deploying malaria nets in the tropics might save even more lives for each dollar spent.[9] MacAskill, Ord, and several others would later dub this data-driven approach to charitable giving "effective altruism," or EA; even before settling on a name, MacAskill and Ord set about evangelizing for the idea, asking friends and colleagues to sign the 10 percent pledge. When they started Giving What We Can, "we had twenty-three members, and most of them were friends of Toby's and mine," MacAskill recalled.[10] MacAskill also cofounded another organization, 80,000 Hours—named for the amount of time spent over the course of a lifetime at a typical forty-hour-a-week job—which focused on providing research-based advice on the best careers to pursue to help other people, including the idea of earning to give.[11]

Meanwhile, MacAskill himself was thriving. In 2015, his first book, *Doing Good Better*, was published, a 272-page argument for the EA approach to charitable giving. That same year, at age twenty-eight, he became an associate professor of philosophy at Oxford, one of the youngest associate philosophy professors in the world at the time.[12] He cofounded CEA, yet another nonprofit organization, which subsumed Giving What We Can and 80,000 Hours into one institutional home.[13] His organizations gave away $9.8 million in grants in 2019 alone, and $2.5 billion in donations had been pledged by over seven thousand people by 2022, the year his second book came out.[14]

That book, *What We Owe the Future*, advocates for something much less straightforward than the benefits of malaria nets in the developing world. MacAskill argues not only for EA but for a specific strain of EA thought known as "longtermism." "Future people count. There could be a lot of them. We can make their lives go better," he writes at the start of the book.[15] "What we do now will affect untold numbers of future people. We need to act wisely."[16] Extending Singer's analogy, MacAskill argues that temporal distance shouldn't be any more relevant to our moral reasoning than spatial distance. "Distance in time is like distance in space. People matter even if they live thousands of miles away. Likewise, they matter even if they live thousands of years hence. . . . Just as the world does not stop at our doorstep or our country's borders, neither does it stop with our generation, or the next."[17]

Most people would agree with MacAskill that we have moral obligations to future generations. Appeals to fight global warming and save fragile ecosystems often invoke a form of this logic, as do arguments for cultural preservation, such as archiving dying languages or preserving ancient artwork. MacAskill is fully on board with such projects. But longtermism implies a great deal more than that. MacAskill's book argues that trying to leave a better world for those who come after us isn't enough—we must also try to ensure that as many people come after us as possible. This is not just about making future generations larger; it's about maximizing the probability that there are as many of those generations as possible, filled to the brim with happy people. MacAskill is thinking about the truly long term. "To illustrate the potential scale of the future, suppose that we only last as long as the typical mammalian species—that is, around one million years. Also assume our population continues at its current size. In that case, there would be eighty trillion people yet to come; future people would outnumber us ten thousand to one."[18] (To put that into perspective, that would mean that currently living humans would be outnumbered by our descendants in the same proportion as the residents of San Francisco are outnumbered by the rest of the world.) We have an obligation to try to make the lives of those humans as good as possible, according to MacAskill. And he claims we are uniquely positioned to do so. "If humanity survives to even a fraction of its potential life span, then, strange as it may seem, we are the ancients: we live at the very beginning of history, in the most distant past. . . . Few people who ever live will have as much power to positively influence the future as we do."[19]

That influence, he claims, extends to whether there will be even more than the aforementioned eighty trillion future humans. About a billion years from now, the Sun's increased heat will vaporize the Earth's oceans, kicking off a runaway greenhouse effect that will make the Earth lethal for water-based life. If our species survives until then, somehow maintaining our present population of about eight billion people over that whole span of time, then there will be about one hundred quadrillion (one hundred million billion, or 10^{17}) future people, twelve million for each human alive today.[20] And if, instead of merely being limited to the Earth's surface, we expand humanity out into space, the numbers of potential future humans become correspondingly astronomical. Over that same billion-year span, spacefaring humans could distribute themselves across the entirety of our Milky Way galaxy, home to at least one hundred billion planets. Even if only 1 percent of those are habitable by humans, that still leaves us with enough room for 10^{26} future humans over the next billion years, if there's an average population of eight billion people per planet at any given time. But that isn't the limit: other planets will have liquid water for far longer than a billion years. And if we can fill the Milky Way, why not the observable universe? If humanity fills the universe to the brim, a burgeoning population across the cosmos until essentially all stars die, the number of future humans could be closer to 10^{40}. That's ten million billion trillion *trillion* people, a one with forty zeroes after it.[21] And this all presumes that our descendants remain human, with our bodies and brains of flesh. If we find a way to transfer human minds into computers, or our primary descendants are themselves conscious AIs, there could be a future filled with unnumbered myriads of electronic life, their silicon circuitry silently traversing the intergalactic voids until the heat death of the universe.

For MacAskill, it's literally the more the merrier. As long as our descendants' happiness outweighs their misery, his logic demands that the greater their numbers, the better the future is. He argues that we should be aiming for the most maximalist of these futures, as best we can. "The future of civilization could be literally astronomical in scale, and if we will achieve a thriving, flourishing society, then it would be of enormous importance to make it so."[22] Yet there are so many things that seem unlikely or impossible in these futuristic visions of the final frontier. The idea that our per-world population will remain at an average of eight billion for hundreds of thousands of years, much less millions or billions of years, is already quite suspect. As MacAskill himself notes, this is an unusual period of growth for the human population, and we're already at an all-time high, which even near-term population forecasts suggest we won't surpass by much. But putting that objection aside, there are far more serious ones to consider. Living in space is phenomenally difficult. There are no good candidates for long-term human habitation in our solar system, and given the distances involved, sending humans to other star systems is extremely unlikely to be anything other than science fiction. Transferring human minds into computers is probably impossible for a variety of good scientific reasons. Conscious AI may be somewhat more likely, but still far from certain—and sending such an AI into space would come with its own set of practical challenges and ethical concerns.

The likelihood of these futures is small, not just because they're scientifically implausible but also because they're rather specific, depending on so many large and small things falling into place, things that we can't know about, like the individual desires and cultural norms of future humans. Yet the specious beauty of longtermism is that the vanishingly small likelihoods of futures that contain vast numbers of humans don't actually matter. In MacAskill's arithmetic, the low probabilities of those futures are outweighed by the number of humans inhabiting them, because the odds of such a future coming to pass, while extraordinarily low, are not zero. Say that there's a one-in-ten-billion chance that humanity will spread out across the accessible portion of the observable universe. And say that there's a one-in-ten-million chance that, if such a future does come to pass, a particular choice you make right now—like donating to a foundation that works on studying possible means of interstellar travel—will measurably help all the humans living in such a future. Then, in total, there's a one-in-10^{17} chance that such a donation will help humans in such a future. (That's around the same odds of winning the Powerball and being killed by lightning on the same day.)

If such a future in space did come to pass, the number of people who would live there would be unthinkably huge. Multiplying the fantastically small odds that you're helping the inhabitants of such a future with your actions now by the even more fantastically huge number of hypothetical people that would inhabit such a future yields the "expected value," the estimate of how many people will be better off, on average, if you were to make that donation. In this case, the answer comes out to about 10^{23} people, many trillions of times more than the number of people currently living—roughly the number of atoms in one breath of air. Thus, given the choice between making such a donation and some other hypothetical course of action that would measurably help every single person alive today, the mathematics of longtermism suggests that making the donation to the space propulsion think tank is the better choice. Helping all eight billion humans alive with a single action is a tall order, but that just deepens the problem: there's no course of action impacting humans here on Earth right now that could possibly compare with the noble mission of helping future humans, because there simply aren't enough humans alive right now to compete with the needs of the hypothetical quadrillions and quintillions of humans in our glorious-yet-improbable science-fictional future.

MacAskill states that his book is merely a case for longtermism, the idea that future people are an important factor in making ethical decisions, rather than a case for what he calls "strong longtermism," the idea that future people matter more than anything else in making ethical decisions. He calls the case for strong longtermism "surprisingly strong" but insists it's not what he's defending.[23] (He does defend it quite vigorously in a separate paper.)[24] But he doesn't have to—the vast numbers of humans in the futures he considers do the work for him. Taken seriously, the moral calculus of his book explicitly demands that such futures must be the overriding consideration in all choices we make about how to effect the most good in the world.

The primary source of the problem here is uncertainty: we don't know what kind of future will come to pass. If we knew, for sure, that there were only two futures possible, one in which humanity goes extinct soon and another where we spread across the stars indefinitely, MacAskill's case would be more compelling. But we don't have that information. This gives the lie to MacAskill's claim that temporal distance is like spatial distance. Distance in space is fundamentally different from distance in time because, while we do have telephones and airplanes, we don't have time machines. We can talk with people from all over the world and even go visit them and ask them what they need. But we can't go to the far future to find out what the people there need from us right now. MacAskill talks about this uncertainty at length in his book, but the drastic conclusions he draws about necessary actions right now, based solely on the possibility of these seemingly outlandish futures, seriously undermines that discussion. He is drawing conclusions that are far too strong based on little more than guesswork about what the distant future could hold. And MacAskill's ability to forecast the future—even in the short term—is seriously questionable. Given far more information than most, he still didn't accurately predict what would happen with Sam, just a few months after *What We Owe the Future* was published.

* * *

The fairly salient problems with longtermism weren't enough to dampen interest in *What We Owe the Future* when it was published in August 2022. A week before the book came out, the *New Yorker* published a ten-thousand-word profile of MacAskill, with a headline dubbing him the "reluctant prophet of effective altruism."[25] The next day, the *New York Times* posted an interview with MacAskill conducted by Ezra Klein.[26] And the day after that, *Time* ran a cover story on effective altruism and MacAskill, concluding that "if the future could be as vast and good as MacAskill thinks, it seems worth trying."[27] Once *What We Owe the Future* actually came out, it landed on the *New York Times* bestseller list almost immediately and stayed there for three weeks straight.[28] Joseph Gordon-Levitt called the book "an optimistic look at the future that moved me to tears"; Stephen Fry said it was "a book of great daring, clarity, insight and imagination."[29] There were a few dissenting voices amid the media hype, but by and large, the launch of *What We Owe the Future* was a resounding success.[30] (Full disclosure: *What We Owe the Future* was published by Basic Books, who also published this book, as well as my first book.)

In and of itself, none of this is particularly remarkable. Plenty of nonfiction books making bad arguments end up with a great deal of media attention and approval. But there's often a reason for it. Books arguing that global warming isn't a big deal get a lot of approving hype from the right-wing media, because it's in their interest to further the narrative that global warming either isn't real or doesn't matter. Effective altruism and longtermism aren't nearly as insidious or destructive as climate denial. But like climate denial, EA has a great deal of corporate money supporting it. Nearly all of that money is coming from the tech industry in the form of donations to the various EA charities. The single largest donor to Effective Ventures—the umbrella nonprofit that now houses CEA, 80,000 Hours, Giving What We Can, and several other EA organizations—is Open Philanthropy, a foundation whose approach to charitable giving is itself based on EA.[31] Open Philanthropy is mostly funded through the fortune of Dustin Moskovitz, one of the cofounders of Facebook, and his wife, Cari Tuna. As of August 2024, Open Philanthropy has donated over $200 million to Effective Ventures and its constituent organizations.[32]

Other nonprofits in the wider EA ecosystem have received even more lavish tech funding. "Existential risk"—threats to humanity as a whole, like the AI apocalypse envisioned by Yudkowsky—is one of the major areas of EA concern. The Future of Life Institute (FLI) is a nonprofit with ties to the EA community that is focused specifically on avoiding such "extreme large-scale risks" from technology.[33] FLI also has strong ties to the tech industry. It was cofounded by Jaan Tallinn, an Estonian tech billionaire who helped develop Skype and Kazaa; Elon Musk has also donated $14 million to the nonprofit.[34] But the overwhelming majority of FLI's money comes from a single source: Vitalik Buterin, the cocreator of the cryptocurrency Ethereum, donated over $650 million (in the form of a different cryptocurrency called Shiba Inu) to FLI in 2021—instantly putting it on a similar financial footing to more well-known and influential think tanks like the Brookings Institution.[35]

EA and longtermism are quite compatible with other causes that have been loci of lavish tech industry interest and funding since well before MacAskill met Ord. The longtermists' idea of a good future requires a phenomenal level of growth—growth in population, in economic productivity, in energy usage, in accessible natural resources. The desire for growth is a general feature of much of capitalism. But the idea of a big future filled with virtually unlimited growth, a future of the specific sort longtermism proffers, has held a great deal of currency in Silicon Valley for decades.[36] The most salient example of this is the concept of a technological singularity, usually referred to as the Singularity.

Believers in the Singularity claim that technological progress has been accelerating and will continue to do so, leading to a singular point where so much change happens so rapidly that the fundamental nature of daily human life will transform beyond all imagination or comprehension. Superintelligent AI and human-machine hybrids will usher in a utopia, end scarcity, and make biomedical discoveries that will allow us to live forever or nearly so. Bounded only by the laws of physics, there will be no practical limit to what a post-Singularity civilization can achieve. According to Ray Kurzweil, the most prominent exponent of the Singularity, the current rate of technological change strongly suggests that the Singularity is coming very soon indeed—no later than twenty years from now, in 2045. "Ultimately, it will affect everything," he claims. "We're going to be able to meet the physical needs of all humans. We're going to expand our minds and exemplify these artistic qualities that we value."[37]

There's little scientific basis for the idea of a Singularity and all the attendant miracles it will supposedly perform. Nonetheless, the idea is astonishingly common in Silicon Valley and across the entire tech industry. Kurzweil isn't some kind of marginal figure. He is a director of engineering at Google, and his books on the Singularity have been bestsellers. "The Singularity is a new religion—and a particularly kooky one at that," said computer scientist and artist Jaron Lanier. "The Singularity is the coming of the Messiah, heaven on Earth, the Armageddon, the end of times. And fanatics always think that the end of time comes in their own lifetime."[38]

This religion is predicated on growth. And the Singularity and longtermism are far from its only manifestations. Rhetoric about the necessity of limitless growth to save the world is commonplace among some of the most prominent tech CEOs. "I believe and I get increasing conviction with every passing year, that Blue Origin, the space company, is the most important work that I'm doing," said Jeff Bezos in 2018. "I'm pursuing this work, because I believe if we don't we will eventually end up with a civilization of stasis, which I find very demoralizing. I don't want my great-grandchildren's great-grandchildren to live in a civilization of stasis." Then he explained the origin of his concerns:

> If you take baseline energy usage globally across the whole world and compound it at just a few percent a year for just a few hundred years, you have to cover the entire surface of the Earth in solar cells. That's the real energy crisis. And it's happening soon. And by soon, I mean within just a few 100 years. We don't actually have that much time. So what can you do? Well, you can have a life of stasis, where you cap how much energy we get to use. . . . Stasis would be very bad, I think. . . . [But] the solar system can easily support a trillion humans. And if we had a trillion humans, we would have a thousand Einsteins and a thousand Mozarts and unlimited, for all practical purposes, resources and solar power unlimited for all practical purposes. That's the world that I want my great-grandchildren's great-grandchildren to live in.[39]

In a talk he gave in 2019, Bezos elaborated on the dangers of stasis: "A life of stasis would be population control combined with energy rationing. That is the stasis world that you live in if you stay [on Earth]."[40] Meanwhile, Elon Musk sees the alternative to growth as more dire than stagnation and rationing: he has framed the quest for space colonies in Manichean terms, a struggle between "the light of consciousness" and the perpetual darkness of extinction.[41] Musk and Bezos aren't alone in such ideas. They're quite common among tech CEOs and venture capitalists, dreaming of a perpetual future of investment opportunities in deep space. The cleanest formulation of this thesis comes from the *Anatomy of Next* podcast, created by Founders Fund, a major tech venture capital (VC) firm. "Human destiny is a binary choice. We can build whole new worlds, around new stars, or we can fade away," Mike Solana, host of the podcast and chief marketing officer of Founders Fund, proclaimed. "This is a podcast about never fading away."[42]

This choice between perpetual growth and the end of humanity is a false dichotomy. Other good futures are possible—and perpetual growth is impossible. Historically, economic growth has always been tied to growth in energy usage. Just as the global GDP has grown, on average, by 3 to 5 percent annually over the past few decades, so has energy usage grown by 3 percent annually, on average, since then. But that can't continue indefinitely. If humanity's energy usage continues to grow by a more modest 2.3 percent per year, then in about four hundred years, we'd reach Earth's limit—we'd be using as much energy as the Sun provides to the entire surface of the Earth annually. (Other nonrenewable energy sources would have long since run out by this point, with the possible exception of uranium. But at that level of energy consumption, the laws of physics guarantee that waste heat would boil the oceans anyhow.)[43] This is what Bezos was referring to: for energy usage to continue to grow past that point, we would have to leave Earth. Yet Bezos seems to have missed a crucial point: while it's true that the energy available on Earth is finite, the energy available in space is just as finite, and just as subject to limits on growth. If growth in humanity's energy usage were to continue at the same rate past the four-hundred-year mark, in 1,350 years we'd be using all the energy produced by the Sun; 1,100 years after that, we'd be using all the energy produced by all the stars in the Milky Way. And about 1,250 years after that, 3,700 years from right now, we'd be using all the energy produced by all the stars in the observable universe.[44] If Bezos believes that ceasing to grow our energy usage must lead to a culture of stagnation, he'd better get used to the idea. Sometime in the next 3,700 years—only about 80 percent of the present age of the Great Pyramid at Giza—humanity must stop growing its energy use. And it's probably going to happen much sooner than that.[45]

Yet the false promise of endless growth as a singular utopia, the only conceivable worthwhile destiny for humanity, shines undimmed by such considerations. Instead, just as with any group that has glimpsed paradise, proponents of this type of future are primarily concerned with imagined fears that could prevent their implausible visions from coming to pass. This allows them to focus on problems they've invented, rather than the real problems that currently face humanity. One of those imaginary problems is the idea that we're not using *enough* energy, a concern Bezos shares with Marc Andreessen, the internet pioneer and venture capitalist. A variation on this point is made by MacAskill in *What We Owe the Future*. He acknowledges that there are ultimate limits to economic growth and energy use, but he believes it's imperative that we continue to grow until we have enough technology to prevent the extinction of humanity.[46] (He's rather vague about what that would look like.) "My concern here is not just with a slowdown in innovation but with a near halt to growth and a plateauing of technological advancement," he writes in a chapter titled "Stagnation." "Stagnation could plausibly be one of the biggest sources of risk of extinction or permanent collapse that we face."[47]

These fears are a sort of twist on a standard fear about ending growth: that it will lead to a recession or worse. ("Worse" can mean fears of war and the end of humanity, concerns that MacAskill and the longtermists share.) But there's another bogeyman that's truly unique to the tech industry. According to Yudkowsky's rationalists, the most pressing challenge in the world is AI alignment: how to ensure that an AGI will have goals and desires that are compatible with those of humanity. The rationalists claim that, without AI alignment, we are hurtling toward an imminent future where the world is in thrall to a superintelligent machine that is not evil per se, but simply does not care about humans and their desires, and has the power to use the raw resources of the Earth—including those that currently compose human bodies—to do whatever it likes.

This apocalyptic vision is the obverse of the Singularity's AI-fueled utopia, and the reasoning behind it is similarly specious—particularly the claim that working to prevent this scenario is the most important problem facing humanity today. Like the tech billionaires' fears of stagnation and fading away, the rationalists' obsession with AI alignment allows them to ignore the real problems of today in favor of the imaginary problems of tomorrow. Longtermism has this problem too, as Peter Singer himself has pointed out. "Viewing current problems—other than our species' extinction—through the lens of 'longtermism' and 'existential risk' can shrink those problems to almost nothing, while providing a rationale for doing almost anything to increase our odds of surviving long enough to spread beyond Earth," he writes. "When taking steps to reduce the risk that we will become extinct, we should focus on means that also further the interests of present and near-future people."[48]

MacAskill and other proponents of effective altruism and longtermism have said that this isn't their view, that effective altruism doesn't say the ends justify the means. "A clear-thinking EA should strongly oppose 'ends justify the means' reasoning," MacAskill wrote.[49] But Singer is right. It's easy to adapt the moral framework of longtermism to fit into the ideology of your choice. Almost anything can be justified in the name of saving the future of civilization.

This is the entire point. As long as billionaires like Elon Musk and Jeff Bezos couch the rationale for their behavior in the apocalyptic terms of longtermism and related ideas—that is, as long as they say that they're doing what they're doing in order to save the future of humanity—then they can cast their critics as enemies of civilization and our species. Musk has done precisely that, quite explicitly: he has said that longtermism "is a close match for my philosophy" and claims that he is simply taking the actions he must take to preserve humanity.[50] "Elon's concept that SpaceX is on this mission to go to Mars as fast as possible and save humanity permeates every part of the company," says Tom Moline, a former SpaceX engineer. "The company justifies casting aside anything that could stand in the way of accomplishing that goal, including worker safety." Moline was fired after making complaints about the workplace at SpaceX. A 2023 Reuters report uncovered over six hundred workplace injuries, including amputations, head wounds, and one death. Most were never reported to OSHA. According to Reuters, SpaceX's "lax safety culture, more than a dozen current and former employees said, stems in part from Musk's disdain for perceived bureaucracy and a belief inside SpaceX that it's leading an urgent quest to create a refuge in space from a dying Earth."[51]

Such monomania makes things simple. Rather than responding in a meaningful way to legitimate criticism—or examining the complicity that they and their companies share in the problems of today—tech billionaires can brush off their critics as lacking sufficient vision to understand their goals. "I have won this lottery, it's a gigantic lottery, and it's called Amazon.com. And I'm using my lottery winnings to push us a little further into space," Jeff Bezos said in 2017. "We need to build reusable rockets, and that is what Blue Origin is dedicated to . . . taking my Amazon lottery winnings and dedicating [them] to [that]. . . . It's a passion, but it's also important."[52] Don't look at the horrifying labor conditions at the local Amazon fulfillment center. Look at the shiny rocket instead. Ignore the problems of this world. Everything will be better in space.

* * *

There is an entire ideology at work here, sprawling and ill-defined. It's fueled by a collection of related desires and shared by a set of influential individuals and communities in the tech industry and the San Francisco Bay Area. These groups—the longtermists, the advocates of the Singularity, the rationalists, and more—share deep connections. They're connected directly by people—there's a great deal of overlap in membership among these groups—and they're connected by a set of common aims and beliefs. Specifically, their ideas have three important features in common, features that go some way toward explaining the popularity of these ideas within these groups and the tech industry at large.

First, these ideas are reductive, in that they make all problems into problems about technology. All the ills of the world will be solved when the Singularity arrives, or when the superintelligent AI solves them for us, or when we go into space. Global warming can be solved with nanotechnology.[53] Illness and death, and all the other problems that come with having a body, can be solved by transferring your mind into a computer. Social problems and political problems—like the problems created by tech companies themselves—are dismissed as irrelevant or unimportant when compared to more urgent problems, like avoiding the creation of an improperly aligned AI or the plight of the hypothetical unborn quadrillions of humans that could live on the other side of the cosmos a billion years from now. It is a philosophy made by carpenters, insisting the entire world is a nail that will yield to their ministrations.

Second, these ideas are profitable, aligning nicely with the bottom line of the tech industry via the promise of perpetual growth. Bezos equates the end of growth with a "civilization of stasis." The long-termists talk about the importance of growing the human species as large as we possibly can, in order to create the largest number of happy lives. The venture capitalists at Founders Fund say an endless future of expansion is the only good one for humanity. The justifications sound noble, but the goal is the same: growth at all costs, growth that will carry corporate profits and billionaires' portfolios up along with it.

Third, and perhaps most importantly, these ideas offer transcendence, allowing adherents to feel they can safely ignore all limitations. Go to space, and you can ignore scarcity of resources, not to mention legal restrictions. Be a longtermist, and you can ignore conventional morality, justifying whatever actions you take by claiming they're necessary to ensure the future safety of humanity. Hasten the Singularity, and you can ignore death itself, or at least assure yourself that you can put it off for a few billion years.

This umbrella of related concepts and philosophies, which I'll call the *ideology of technological salvation*, sits at the core of the worldview held by many venture capitalists, executives, and other "thought leaders" within the tech industry. This ideology promises a glorious future: technological progress, unchecked. Align the AI, avert the apocalypse, and technology will handle the rest. Humanity will expand across the cosmos, exploiting ever-increasing stores of natural resources. All limits to economic growth and energy usage will melt away. The AI will extend our lifespans by a trillion-fold, merging with us or uploading our minds into its silicon paradise. The messy details of sectarian conflict, political struggles, identity politics, and inequality of all kinds will be rendered irrelevant. Working to hasten this utopia by optimizing the shit out of things is the greatest possible good.

This future (or set of futures) doesn't work. Picking just one example, there's no good, scientifically based argument that the future of humanity will be any more secure if we build a settlement on Mars—doing so might actually make the future less secure—nor does the technology to build such a settlement exist right now. Indeed, there's good reason to think that it's effectively impossible to put a self-sustaining human civilization on Mars (at least, not without radically restructuring human biology, which may also turn out to be impossible). The radiation levels are too high, the gravity is too low, there's no air, and the dirt is made of poison. There are many other problems with this idea, and it's one of the simpler ones involved in these visions of the future. Interstellar travel makes going to Mars look like running a quick errand after work. Differences between computer architecture and human neurophysiology (not to mention the difficulty of in vivo, atomic-scale brain scanning) make transferring a human mind to a computer a dubious prospect at best. The list of problems goes on. But these problems are no obstacle to the ideology of technological salvation, which garbs itself in the raiment of science, understanding the power held by the imprimatur of scientific truth. Science is invoked as justification for the claims made by the ideology on the one hand, while actual scientific concerns about the plausibility of these claims are dismissed by the other hand, just as easily as all other limitations are dismissed.

EA and longtermism fit neatly into this picture. It doesn't matter that we're ignoring the plight of island nations in the Pacific as they're swallowed by rising oceans and battered by global-warming-fueled storms, and it doesn't matter that there's good scientific reasons to doubt that humanity can spread across the cosmos. There just needs to be a one-in-a-quintillion chance that such a future could come to pass, one where decillions of people who don't yet exist will live on planets we haven't discovered, using technology that might never be possible to build.

Despite his insistence that the ends don't always justify the means for effective altruists, at times MacAskill has come close to endorsing a version of EA that sounds remarkably like a new gospel of wealth, giving the wealthy a patina of moral rectitude, especially when he's talking about earning to give. "Obviously there's some worry that you're disconnected and lose your values, but I'm coming around to the idea that the rate of doing that via earning to give is no worse than the rate of doing that through direct impact," he says. "If you're earning to give, you're in a cushy lifestyle—you're giving away 50 percent, but you're still on a nice salary—working with very smart people, and you know that the impact you're having is absolutely huge because you're able to donate to these very well-evidenced charities."[54]

Earn to give is predicated on the idea that the source of the money just isn't as important as the causes it goes to. In a paper laying out the moral case for earning to give, MacAskill writes that it is often "ethically preferable to pursue philanthropy through a higher paid but morally controversial career," like "working for a petrochemical company, working for a company involved in the arms industry, and some careers within finance, such as those that involve speculating on wheat, thereby increasing price volatility and disrupting the livelihoods of the global poor."[55] Hence EA's cozy connections with tech billionaires, which MacAskill and other leaders of the movement have cultivated. His friend Sam, meanwhile, became one of those billionaires. And when Sam set up a charitable giving fund, he offered MacAskill a leadership position in it. Sam's money was coming from cryptocurrency, a notoriously volatile and environmentally destructive set of financial instruments. He assured MacAskill that he'd purchased carbon offsets and was pushing to decrease the carbon footprint of the crypto industry overall.[56] So MacAskill accepted Sam's offer, and helped to dole out his fortune. But as it turned out, MacAskill was right: sometimes, the ends don't justify the means.

* * *

While Sam was working at the Centre for Effective Altruism, he started thinking about ways to earn very large amounts of money so he could make a bigger impact on the world (or so he claimed later).[57] He found one in the form of Bitcoin arbitrage—taking advantage of local differences in the exchange rates between Bitcoin and normal "fiat" currencies, like the US dollar and the Japanese yen. Sam knew that in theory, it was possible to make money by buying bitcoin in dollars, selling it in yen, and then exchanging the yen back into dollars. In practice, Japanese banking regulations intended to prevent money laundering and other criminal activities made this set of trades next to impossible to perform.

What happened next isn't entirely clear. According to Sam, he found a small, rural Japanese bank willing to process the transactions he wanted to make. One of Sam's EA connections in Japan set up an account there. Starting with $50,000, Sam's new company, Alameda Research, started making trades, shuttling Bitcoin back and forth between the United States and Japan with a 10 percent return every day. Once he had proven he could do this consistently, Sam used his EA connections again. He started hiring; his first fifteen employees came from the EA community. "This thing couldn't have taken off without EA," said Nishad Singh, one of Sam's early hires. "All the employees, all the funding—everything was EA to start with."[58] Sam secured a $110 million loan from Jaan Tallinn (at a blisteringly high 43 percent interest rate) that served as much of Alameda's initial capital. Tallinn called in most of the loan just a few months later, but by then Alameda was bringing in more than enough to cover it.[59] Sam used his new fortune to build an empire.

In 2019, Sam launched a new cryptocurrency trading platform. If it was successful, as Sam wagered it would be, it would allow him and his EA colleagues at Alameda to profit off transaction fees and investment of the capital they'd have on hand, like an investment bank does with its clients—earning to give on a massive scale. The catch was that cryptocurrency trading is ethically questionable: among other reasons, the phenomenally complex computations involved are massively energy intensive for a computer to perform, and most of that energy is produced using fossil fuels. In 2022, the worldwide energy consumption due to cryptocurrency activity was estimated at 120–240 billion kilowatt-hours, more electrical power than all of Australia uses in a year, approaching 1 percent of all electricity usage worldwide. The carbon footprint associated with crypto is comparably huge, with emissions of about 140 million metric tons of carbon per year, more than annual emissions from Austria, Norway, and Portugal combined.[60] In exchange for that hit to the environment, the economy gets an unregulated financial instrument more cumbersome and less useful than normal currency. But Sam was certainly correct that he could make a lot of money: within two years of starting his new cryptocurrency trading firm, Sam had turned his millions into billions. He recruited friends from his time at Jane Street, and from the EA community, to join him in the upper echelons of his company. He tapped MacAskill to help run the charitable foundation he started to give away his new wealth. By 2021, Sam's company was the third-largest crypto trading platform in the world.[61] The company's directors were entertaining celebrities and investors at their new corporate headquarters in the Bahamas. Sam had a net worth of over $20 billion, making him the richest person under thirty in the world. He even landed on the cover of *Fortune* magazine in August 2022, the same month *What We Owe the Future* was published, with a headline asking if Sam Bankman-Fried was "the next Warren Buffett."

You probably already know the rest of the story. Sam Bankman-Fried's cryptocurrency exchange, FTX, imploded in November 2022. He and his lieutenants at FTX used customers' private account funds to cover trades made by Alameda Research, which Bankman-Fried (aka SBF) and his EA gang were still running as a hedge fund alongside the trading firm. When Alameda's funds crashed with the crypto market in fall 2022, FTX's customers were left holding the bag, with an estimated $8 billion in customer funds gambled away on bad bets and poured into luxury real estate, political donations, and other extravagant purchases by Bankman-Fried and company. SBF was arrested in the Bahamas in December 2022 and extradited to the United States; he was later convicted of securities fraud, wire fraud, money laundering, and several other charges. He was ultimately sentenced to twenty-five years in federal prison.[62]

The timing of Bankman-Fried's fall couldn't have been worse for MacAskill and the longtermists. Just a few months after the launch of *What We Owe the Future*, while the good press that it had generated was still coming in, SBF's fall hit the news. Coverage of the fraudulent cryptocurrency boy-king was everywhere, and a good number of the longer articles went into the unusual moral philosophy purportedly underlying his career. At the same time SBF was committing large-scale financial fraud, he had been donating lavishly to a host of EA organizations. The FTX Future Fund, an arm of FTX's philanthropic organization with a team composed of MacAskill and other effective altruism luminaries, had given grants to typical EA and longtermist causes such as AI alignment research and pandemic preparedness. They gave to MacAskill's own organizations, including nearly $27 million to Effective Ventures.[63] They also gave millions of dollars to other longtermist and rationalist organizations and recommended investments to promote longtermist and rationalist ideas, including $400,000 "to support the creation of animated videos on topics related to rationality and effective altruism to explain these topics for a broader audience."[64] SBF's family foundation, Building a Stronger Future, had also made donations to news organizations focused on effective altruist subjects, including $200,000 to the news website *Vox*—which already had a section, Future Perfect, dedicated to journalism from an EA perspective—specifically for reporting on "technological and innovation bottlenecks that hamper human progress," and millions of dollars to other outlets like ProPublica and *Semafor*.[65] (The grants to Effective Ventures, *Vox*, ProPublica, and *Semafor* have since been returned to the FTX estate as part of the ongoing bankruptcy proceedings.) Now, thanks to SBF, the EA community had finally gotten its wish. Effective altruism had broken into mainstream awareness—in the worst possible way. "How Effective Altruism Let Sam Bankman-Fried Happen," read one *Vox* headline in the immediate aftermath of SBF's fall.[66] The *New York Times*, the *Atlantic*, the *New Yorker*, the *Washington Post*, and many others ran similar headlines.[67]

Some news reports on the implosion of FTX also picked up on stranger facets of SBF and his circle, things that didn't quite make sense on the face of it. For example, Caroline Ellison, SBF's ex-girlfriend, former colleague at Jane Street, and former CEO of Alameda Research, seemingly ran a strange Tumblr account. Media reports on the account, titled "worldoptimization," collected extremely compelling evidence that Ellison was behind it (the author of the Tumblr shared many traits with Ellison, the account linked to her Twitter, and SBF said it was hers).[68] The media also focused on the fact that she posted approvingly about the racist pseudoscience of human biodiversity, and that she said she used to believe "the sexual revolution was a mistake" and that "women are better suited to being homemakers and rearing children than doing Careers."[69] But there were other posts, too, ones that made even less sense to most observers. On a list of "~cute boy things~" she included "controlling most major world governments" and "being responsible for many important inventions and scientific discoveries," following that list with a set of affirmations:

- if you are a boy who is driven to succeed at ambitious goals *you are valid*

- if you are a boy who arrives at opinions through logical reasoning *you are valid*
- if you are a boy with the confidence to advocate for unconventional ideas and take actions based on them *you are valid*[70]

Those quotes from Ellison's alleged Tumblr appear to be indirect references to a foundational text in the rationalist community, and the name of the Tumblr itself, "worldoptimization," is a direct reference to it.[71] While effective altruism is theoretically independent of rationalism, the two groups have so much overlap in people—so many effective altruists are rationalists and vice versa—that in practice, it's not possible to fully parse the behaviors and motivations of the two groups independently of each other. Taken in context, Ellison's Tumblr posts seem less bizarre and more chilling. She and others at FTX seemed to believe that, in order to save the world, they needed to accumulate wealth and power to steer the future of humanity by sheer economic and political force, whether the rest of us liked their ideas or not.

But whether SBF actually believed he was saving the world, rather than merely accumulating wealth and power, is itself questionable. In the summer of 2022, he said to journalist Kelsey Piper (herself an effective altruist who works at Future Perfect, the EA-affiliated section of *Vox*), "There are a lot of complicated but important second-order harms that come if your core business is bad for the world, in terms of... your ability to work with partners in your philanthropic efforts." But that November, when FTX was falling down around him, he had another conversation with Piper (this time over Twitter DMs), and she asked him whether that was the real answer or just what he was supposed to say. "Man all the dumb shit I said," he wrote. "It's not true, not really. . . . I feel bad for those who get fucked by it, by this dumb game we woke westerners play where we say all the right shiboleths [*sic*] and so everyone likes us."[72] SBF later told the *New York Times* that he had been referring to greenwashing and similarly hollow corporate PR campaigns; he also claimed that he'd "stupidly forgot" that Piper was a journalist.[73]

But whether or not the ultra-wealthy tech elite actually believe in the ethical justifications offered by longtermism and the futures promised by technological salvation, such ideas can serve as convenient forms of public relations. Regarding Elon Musk's plans to settle people on Mars, "I don't doubt that he wants to do that and that he thinks that's an exciting idea," Lucianne Walkowicz, astronomer and cofounder of the JustSpace Alliance, tells me. But, they add, "it sounds a whole lot better than, 'We'd like to have more NASA contracts, please.' . . . And I think sometimes the 'Why do people talk about [wanting to go to Mars]?' actually is not tied to wanting to go at all. It's tied to a kind of story crafting about what their Earth-based projects are really doing."[74]

Even if Silicon Valley billionaires are deploying the language of technological salvation and spinning out its visions of the future just to garner goodwill from the public, that only works because other people really do believe in those ideas, or at least find them plausible. So understanding the ideology of technological salvation isn't just about understanding the motivations of the ultra-wealthy. It's also a crucial step in deflating their power. Technological salvation is being used as an excuse to steer society in a dangerous direction, in the service of an impossible future. Breaking free of these visions means understanding them. For the tech elite, these are visions of transcendence, of escape. But they hold no promise of escape for the rest of us, only nightmares closing in. To wake from the dream, we must first understand its shape. And the best place to start with that is the purest expression of this ideology of perpetual growth, that fantastical vision of a perfect, unstoppable, inevitable technological utopia: the Singularity.

2

MACHINES OF LOVING GRACE

Ray Kurzweil is pretty sure his dad isn't going to stay dead. "I have 50 boxes of his things at home, his letters and music and bills and doctoral thesis," Kurzweil, an inventor, author, and futurist, says. His plan for paternal resurrection involves feeding those documents into a computer, along with more visceral sources of information about his father. "We can find some of his DNA around his grave site," Kurzweil says. An "AI will send down some nanobots and get some bone or teeth and extract some DNA and put it all together. Then they'll get some information from my brain and anyone else who still remembers him. . . . [They'll] just send nanobots into my brain and reconstruct my recollections and memories."[1] Armed with all this data, the AI will create a program capable of reproducing Kurzweil's father's behavior, responding to new situations as his father would have and even holding a conversation. "I will be able to talk to this re-creation. . . . Ultimately, it will be so realistic it will be like talking to my father," Kurzweil claims. "You can certainly argue that, philosophically, that is not your father," he continues, "but I can actually make a strong case that it would be more like my father than my father would be, were he to live."[2]

Kurzweil, in his midseventies, speaks in a measured and faintly gravelly New York accent, sounding rather matter-of-fact while making his surprising claims. But behind these claims is a deep and broad technological expertise: Kurzweil has been named as an inventor on dozens of patents over the course of his long career. Those inventions range from new kinds of electronic synthesizers to early text recognition software to assistive devices for the blind. He first started working with computers in 1960, at the age of twelve, building a "computer-like device" for a science fair at his junior high school, then moving on to IBM mainframes and coding in Fortran with the help of his uncle, an engineer at Bell Labs.[3] He often discussed his ideas with his father, a conductor and music educator who earned a PhD in musicology from the University of Vienna before fleeing Austria to escape the Nazis in 1938.[4] "We talked a lot about the nature of music and mathematical structure, and the fact that computers and music had natural affinity to each other," Kurzweil told *Rolling Stone* in 2009. "He said, 'Someday you'll get involved in creating synthetic music, using computers.' He recognized that eventually computers could do a better job." But in 1970, when Kurzweil was twenty-two—the same year he graduated from MIT—his father died of a heart attack at the age of fifty-seven. Mourning his father, Kurzweil spent some time after college thinking about a way to bring the dead back to life, approaching the ancient subject with an engineer's eye. "A person is a mind file. A person is a software program, a very profound one, and we have no backup. So when our hardware dies, our software dies with it," he told *Rolling Stone*. "I've made an issue of overcoming death. . . . And the strongest experience I've had with death is as a tragedy."[5]

That tragedy, according to Kurzweil, was the original impetus behind his interest in exploring the possibilities that will come with technology in the near future, an exploration that has led him to his certainty that his father will be back, soon. But nobody knows how to collect memories from a living human brain, nor does anyone know how to build an AI that could synthesize all this information and spit out a chatbot that perfectly simulates conversations with a now-dead human. Yet Kurzweil has a track record of accurately predicting the future of technology, especially AI. In 1990, he claimed that a computer would beat a human at chess by 2000; it happened in 1997.[6] And in 1999, he accurately predicted that airborne drones would be commonplace in military conflicts by 2009. Bill Gates once said that nobody was better than Kurzweil at predicting the future of artificial intelligence.[7] In 2010, Kurzweil put his own accuracy at 86 percent (though some independent estimates of his accuracy have been much lower). This may explain his quiet confidence in his own forecasts, even when they seem outrageous.[8] And his forecasts for the next twenty years are certainly not pedestrian. He has repeatedly claimed that by 2029, an AI will be able to "do everything that any human being can do."[9] By 2045, he says, "we will multiply our effective intelligence a billion fold by merging with the intelligence we have created"—an event Kurzweil and others call "the Singularity."[10]

The Singularity gets its name from the point where a mathematical function breaks down, usually because it's shot off toward infinity. Here, the function is the rate of progress itself. Specifically, Kurzweil claims there is a "law of accelerating returns," an exponential trend in technological progress proving that the Singularity is not just likely but inevitable, and coming soon. "People intuitively assume that the current rate of progress will continue for future periods. Even for those who have been around long enough to experience how the pace of change increases over time, unexamined intuition leaves one with the impression that change occurs at the same rate that we have experienced most recently," he writes. "But a serious assessment of the history of technology reveals that technological change is exponential."[11]

That exponential change, he says, will radically transform human life past the point where we can recognize it—and that will happen soon. Hence the title of one of Kurzweil's books: *The Singularity Is Near*.

"We won't experience 100 years of progress in the 21st century—it will be more like 20,000 years of progress," Kurzweil claims.[12] After the Singularity, our current society will be a distant memory, just as the preagricultural societies of Ice Age humans are to us now. There won't be poverty, disease, or want; all human desires will be satisfied immediately by the ubiquitous and nearly all-powerful machines, enhancing our lives and ourselves. "We're going to be funnier, we're going to be better at music. We're going to be sexier," Kurzweil says. "We're really going to exemplify all the things that we value in humans to a greater degree."[13] And any of us alive then will "live as long as we want," Kurzweil wrote in *The Singularity Is Near*.[14] He claimed in 2009 that by 2024 technology would be advanced enough to rejuvenate him to a biological age of forty; this doesn't seem to have worked out.[15] (I had hoped to ask him about this myself, but he declined to be interviewed for this book.) That hasn't discouraged him. "We're going to get to a point where we have longevity 'escape velocity,'" he said at South by Southwest in 2024. "By 2029, if you're diligent, you'll use up a year of your longevity with the year passing, but you'll get back a full year, and past 2029, you'll get back more than a year. So you'll actually go backwards in time [i.e., get younger as time passes]." When asked if anyone in the audience that day would live for five hundred years, he didn't hesitate: "Absolutely. I mean, if you're going to be alive in five years—and I imagine all of you will be alive in five years," he said.[16] And if Kurzweil himself doesn't make it to 2029, he's still relatively sanguine about his own prospects for immortality. Dying before then "would be a setback," he admits. But in that worst-case scenario, he says he'd just leave instructions for the post-Singularity robots to sort through his belongings and scan his frozen corpse—bringing a version of himself back to life, just like his father.[17]

* * *

To understand exactly why Ray Kurzweil thinks he can resurrect his father and that anyone alive in 2029 will still be alive in 2529—and, more generally, to understand what the Singularity is—requires a careful look at what Kurzweil was talking about regarding linear growth and exponential growth. Something grows linearly when you add a constant amount to it over and over again. Make a dollar a day, and the total amount you've made will just increase by $1 each time: $1, $2, $3, and so on. Crucially, with linear growth, it doesn't matter how much you already have: on the tenth day, you still only make $1, and your balance goes from $9 to $10. Not so with exponential growth. There, instead of adding a constant amount, you multiply by a constant amount, increasing your total by a fixed percentage of what you already have rather than adding a fixed sum. Instead of making a dollar a day, you deposit a dollar on the first day into an account that earns 100 percent interest each day, doubling your money daily. Now, the amount you already have matters quite a bit. On the second day, you only get one additional dollar. But the next day you get two more dollars, and the day after that, four. With exponential growth, having more leads to getting more—and that, in turn, leads you to get even more. That compounding interest means that exponential growth can become surprisingly fast in short order. After the third day, you have $4 in your bank account, while your friend who stuck with linear growth has $3—you're not doing much better. But after the eleventh day, your friend has just $11, while you have over $1,000 in your account. At the end of the month, on the thirty-first day, your friend has $31, while you've just become a billionaire; at the end of the next month, you'll have over $1 quintillion, more than five thousand times the total world GDP in 2021.

According to Kurzweil, technology works this way too. The most famous example of an exponential trend in technology is probably

Moore's law, named after the microchip pioneer and cofounder of Intel, Gordon Moore. In 1965, Moore famously observed that the number of transistors that could fit on a silicon wafer was doubling every year. Ten years later, he revised this estimate to a doubling every two years; as others soon pointed out, this implied that the speed of the fastest computer chips (made of silicon wafers filled with transistors) would double roughly every eighteen months. This trend continued for decades after Moore's pronouncement, past the end of the twentieth century.[18]

To Kurzweil, this is a manifestation of a more general feature of human technology. "Technology, like any evolutionary process, builds on itself," he writes. "Humans are now working with increasingly advanced technology to create new generations of technology."[19] Because our tools enable us to build better tools, which in turn enable still better tools, Kurzweil says it makes sense for technological progress to grow exponentially. Thousands of years ago, the development of pottery furnaces enabled smelting, which enabled metalworking, which ultimately allowed bronze and then steel to replace stone in tools and weapons. More recently, better and better computers have made it easier to design almost anything—including the next generation of computers themselves. Hence Kurzweil's law of accelerating returns, which he thinks is a fundamental feature of human technology.

This is not a merely theoretical argument from Kurzweil. He has collected a set of major technological milestones over the course of human history, and has found that they are coming more and more quickly, matching an exponential trend. It took five thousand years to go from the horse-drawn cart to the car, but less than a century to go from the car to landing on the Moon; there were at least two thousand years between homing pigeons and the telegraph, but only about 130 years between the telegraph and the beginning of the internet.

Nor does Kurzweil think that the law of accelerating returns is limited to technology. According to him, exponential trends can be traced back through the deep time of evolution as well. "Exponential growth is a feature of any evolutionary process," he writes.[20] "The evolution of life-forms required billions of years for its first steps (primitive cells, DNA), and then progress accelerated. During the Cambrian explosion, major paradigm shifts took only tens of millions of years. Later, humanoids developed over a period of millions of years, and Homo sapiens over a period of only hundreds of thousands of years."[21] (See Figure 2.1.) And, Kurzweil says, he's not the only one seeing this. He found the same trend when using milestones picked out by fifteen different sources on the history of life, the universe, and technology.[22] Our technology, according to Kurzweil, is merely a continuation of the exponential trend that started with the first self-replicating molecules arising on Earth billions of years ago.[23]

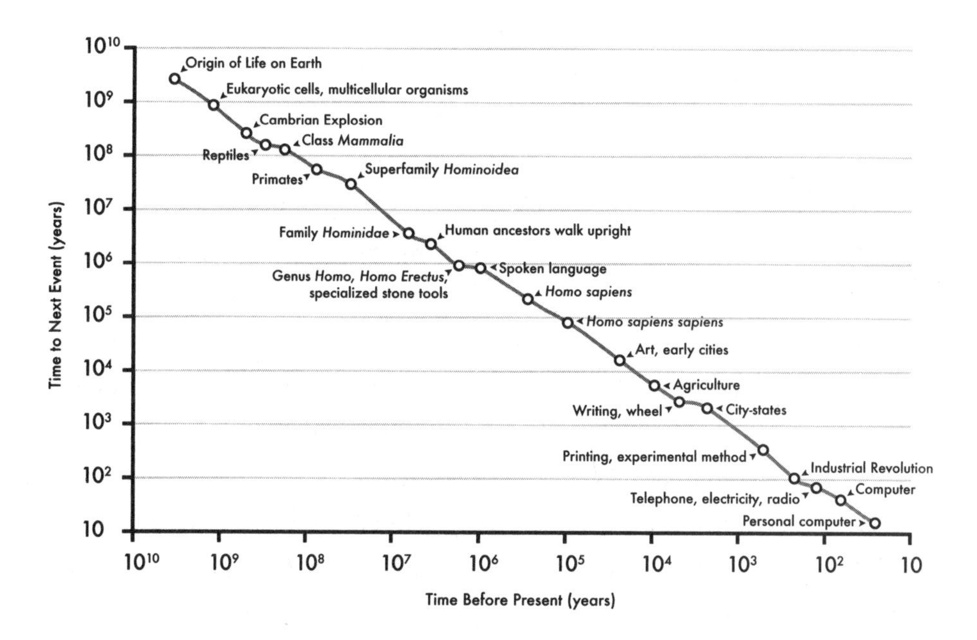

Figure 2.1: Kurzweil's claimed exponential trend in evolutionary and technological history. This actually depicts a linear countdown to the present moment on a logarithmic scale, not an exponential trend.

The next step in this trend is artificial intelligence. By this, Kurzweil doesn't mean existing AI systems and machine learning (ML) algorithms. He has something much more powerful—and profound—in mind. Kurzweil calls it "strong" artificial intelligence or "artificial general intelligence," which he defines as "artificial intelligence that exceeds human intelligence."[24] That definition is more than a little vague, but Kurzweil isn't bothered by that. Once humans manage to build an AGI, Kurzweil says, we can run it far faster than humans can think, enabling it to perform difficult intellectual tasks much more quickly than a human could.

One of those tasks, naturally, would be the design of the next AGI. And this is the mechanism by which Kurzweil and other "singularitarians" see the Singularity coming to fruition. Say that first AGI, built by humans, took thirty years to create. Building an even better, smarter one might take those same humans more like sixty years. But with an AGI in hand, it would be easy to speed that up: just ask the AGI to design it, and give it immense computational resources. Then that first AGI could do work that would take humans sixty years in just a year, or even a few days. And once that smarter AGI is built, it would be able to design an even smarter one than itself even faster, and that one could do the same, and so on, leading to sudden, runaway growth—exponential, or perhaps even faster—in the intelligence of AGIs. So the first AGI built by humans could also be the last: the rest, designed and built by successive generations of machines at an ever accelerating rate, would rapidly ascend to unfathomable heights of superintelligent thought, as far above our abilities as we are above ants or bacteria. "Once a computer achieves a human level of intelligence, it will necessarily soar past it," Kurzweil writes.[25] "This cycle of machine intelligence's iteratively improving its own design will become faster and faster."[26] When that happens, he says, we will be hopelessly left behind—unless we go along for the ride. The Singularity "will result from the merger of the vast knowledge embedded in our own brains with the vastly greater capacity, speed, and knowledge-sharing ability of our technology," Kurzweil writes. "[This] will enable our human-machine civilization to transcend the human brain's limitations of a mere hundred trillion extremely slow connections. . . . We will become vastly smarter as we merge with our technology."[27] And as the law of accelerating returns continues to expand the computational power of humanity, "ultimately the nonbiological portion of our intelligence will predominate."[28]

Kurzweil goes on to point out that this is nothing really new. We already live in close connection with our technology. Our phones sit snugly in our pockets, our watches are strapped to our wrists, pacemakers live in many of our hearts, and increasingly sophisticated prostheses replace our lost limbs. To Kurzweil, letting our brains merge with our machines is simply the next step in a trend that started long ago with the wheel, the lever, and fire. And after the Singularity arrives, he says, our human-machine civilization will proceed with the final step in that trend, fulfilling our ultimate destiny: changing the structure of the universe itself.

* * *

In 1962, an odd little book of essays was published under the title *The Scientist Speculates: An Anthology of Partly-Baked Ideas*. "A partly-baked idea or **pbi** is either a speculation, a question of some novelty, a suggestion for a novel experiment, a stimulating analogy, or (rarely) a classification," the editor, I. J. "Jack" Good, wrote in the first essay in the book. "The bakedness of an idea should be judged by its potential value, the chance that it can be completely baked, its originality, interest, stimulation, conciseness, lucidity and liveliness. It is often better to be stimulating and wrong than boring and right."[29] The rest of the book comprises over 120 short essays with titles like "Deliberate Misplints," "Robotic Croquet," "Winking at Computers," "Precognition and Reversed Causality," and "A Theory Which Is Impossible to Believe if True"—the first and last of these written by Good, who described himself as the "perpetrator" of the book.[30] Other contributors included illustrious scientists (Eugene Wigner, Michael Polanyi, John Maynard Smith), authors (Arthur Koestler, Isaac Asimov, Arthur C. Clarke), at least one famous crackpot (Herbert Dingle), and a few people who wished to remain anonymous. One or two of the essays went on to be classics in their fields; most of the rest were quickly forgotten.[31]

By the time that book was published, Good had already been working with computing machines for twenty years. During World War II, the young Good joined the code-breaking team at Bletchley Park upon finishing his PhD in mathematics at Cambridge. Shortly after arriving, the leader of the team, Alan Turing, caught Good napping on the job rather than breaking Nazi codes. But Good quickly found his way into Turing's favor by solving a code-breaking problem that Turing himself hadn't managed to crack. "I thought that I had tried that," Turing sheepishly said to Good upon his success.[32]

After the war, Good bounced between academia, government, and think tanks, ultimately landing at Virginia Tech a few years after *The Scientist Speculates* was published and staying there for the rest of his long life. But throughout his career, Good had been intrigued by the possibilities that computers—and especially artificial intelligence—might hold for humanity. In a book review Good wrote in 1951, he theorized that a "superhuman" machine thinking for itself could serve as "a modern oracle." "The threshold between a machine which was the intellectual inferior or superior of a man would probably be reached if the machine could do its own programming," he wrote.[33] He returned to this theme several times over the years, including in one of his numerous contributions to *The Scientist Speculates*, titled "The Social Implications of Artificial Intelligence," where he wrote that once a computer is built that is itself able to design a better computer, "there would unquestionably be an explosive development in science."

He even provided a prediction of when he thought this would come to pass: he said it was likely to happen by 1978 and would cost around $500 million, give or take a power of ten. Good called that "cheap at the price."[34] He elaborated on this idea in a longer article in 1965, "Speculations Concerning the First Ultraintelligent Machine":

> The survival of man depends on the early construction of an ultraintelligent machine. . . . Let an ultraintelligent machine be defined as a machine that can far surpass all the intellectual activities of any man however clever. Since the design of machines is one of these intellectual activities, an ultra-intelligent machine could design even better machines; there would then unquestionably be an "intelligence explosion," and the intelligence of man would be left far behind. Thus the first ultraintelligent machine is the *last* invention that man need ever make, provided that the machine is docile enough to tell us how to keep it under control.[35]

Good wasn't the only one to have this idea. The concept of a super-intelligent machine had been entertained by his old boss, Turing, as well as AI pioneers Ray Solomonoff and Marvin Minsky (the latter of whom also contributed two essays to *The Scientist Speculates* and later mentored a young student at MIT named Ray Kurzweil). More generally, the concept of technological acceleration leading to a singularity was arguably developed by John von Neumann, the legendary mathematician and physicist. Von Neumann worked with early computing machines on the Manhattan Project and created the basic architecture that all electronic computers still use today. He also wrote one of the first modern treatises on self-replicating machines. After von Neumann's untimely death at fifty-three in 1957, his friend Stanislaw Ulam—also a renowned mathematician and physicist—wrote a lengthy obituary for him. Von Neumann's "conversations with friends on scientific subjects could last for hours," Ulam wrote. "One conversation centered on the ever accelerating progress of technology and changes in the mode of human life, which gives the appearance of approaching some essential singularity in the history of the race beyond which human affairs, as we know them, could not continue."[36]

Von Neumann himself never wrote about this idea. But decades after von Neumann's death, Vernor Vinge took the concept—and the name "singularity"—and ran with it. Vinge was a computer scientist and mathematician by training, but he was best known for his science fiction novels, especially the Hugo Award–winning books *A Fire Upon the Deep* and its prequel *A Deepness in the Sky*. But it was an earlier novel, *Marooned in Realtime*, published in 1986, where Vinge introduced the idea of the Singularity to the readers of his fiction.[37] In *Marooned*, Vinge used a post-Singularity civilization as a plot device: the characters awake in the far future after being placed in stasis, finding that human civilization has seemingly vanished or transcended to a new plane of existence, with a technological Singularity fingered as the likely culprit. In a curious afterword to the book, Vinge explained that he thought this was not merely a fictional conceit. As the year 2000 approached, it was "an ironic accident of the calendar that all this religious interest in transcendental events should be mixed with the objective evidence that we're falling into a technological singularity. . . . It's *you*, not Della and Wil [the protagonists of *Marooned*], who will understand the Singularity in the only possible way—by living through it."[38] Expanding on these ideas in a 1993 article titled "The Coming Technological Singularity," Vinge cited von Neumann and especially Good as forerunners to his ideas. "I believe that the creation of greater than human intelligence will occur during the next thirty years," wrote Vinge, rehearsing the same argument Kurzweil would make more than ten years later in *The Singularity Is Near*.

> When greater-than-human intelligence drives progress, that progress will be much more rapid. In fact, there seems no reason why progress itself would not involve the creation of still more intelligent entities—on a still-shorter time scale. . . . From the human point of view this change will be a throwing away of all the previous rules, perhaps in the blink of an eye, an exponential runaway beyond any hope of control. Developments that before were thought might only happen in "a million years" (if ever) will likely happen in the next century.[39]

Vinge was mostly unwilling to guess about the detailed nature of post-Singularity technology or civilization. "If I get pushed hard about questions about what the Singularity is going to be like," he said, "my most common retreat is to say, 'Why do you think I called it the Singularity?'"[40]

Nonetheless, the prospect of a tech-fueled civilizational apotheosis meshed well with the ideological foundations of the futurist groups already swirling around Silicon Valley in the 1980s and '90s—and they largely didn't share Vinge's reluctance to speculate about what the Singularity might bring. "We see this need for transcendence deeply built into humanity," said the futurist Max More in 1994. "It seems to be something inherent in us that we want to move beyond what we see as our limits. In the past we haven't had the technology to do that, and right now we're in this difficult period where we don't quite have the technology yet, but we can see it coming. . . . I enjoy being human but I am not content."[41] More (born Max O'Connor) was one of the founders of the "Extropian" movement, named "as a metaphorical opposite" of entropy, the physical measure of disorder and decay, which the Extropians declared "the supreme enemy of human hope."[42] More, along with Tom Morrow (born Tom Bell) started *Extropy* magazine (originally subtitled *Vaccine for Future Shock*) in 1988. The first issue declared that the magazine would cover subjects like "artificial intelligence, cognitive science and neuroscience, intelligence-increase technologies, life extension, cryonics and biostasis, nanotechnology, spontaneous orders, space colonization, economics and politics (especially libertarian), science fiction and reviews of science fiction, intelligent use of psychochemicals, extropic psychology, mindfucking, extropic moral and amoral theories, extropic scientific developments, memetics, [and] aesthetics."[43]

More and Morrow also founded the Extropy Institute in 1992, which led in short order to something ultimately more influential than their magazine: an email list for Extropians. "In the mid-nineties, many got [their] first exposure to transhumanist views from the Extropy Institute's listserv," wrote the philosopher Nick Bostrom in 2005.[44] Bostrom himself was one of those people and was quite active on that list in the 1990s, when he was a graduate student in philosophy. In 1998, Bostrom cofounded the World Transhumanist Association, an organization dedicated to the "feasibility of redesigning the human condition, including such parameters as the inevitability of aging, limitations on human and artificial intellects, unchosen psychology, suffering, and our confinement to the planet Earth," as the group's Transhumanist Declaration said. "In planning for the future, it is mandatory to take into account the prospect of dramatic progress in technological capabilities," they continued, echoing Vinge. "We need to create forums where people can rationally debate what needs to be done, and a social order where responsible decisions can be implemented."[45]

Most of those debates happened on the internet, but there were also Extropian parties, conferences, and at least one student organization, at MIT.[46] The Extropians and their intellectual compatriots were profiled in magazines like *Wired* and in several books.[47] By the end of the 1990s, Extropians and transhumanists like Hans Moravec and Ray Kurzweil had published nonfiction books of their own, expounding on the imminent approach of AGI and other transformative technologies as the Singularity purportedly loomed.[48] Kurzweil even quoted More's "Principles of Extropy" in *The Singularity Is Near*.[49]

But as Vinge pointed out in his 1993 article, the idea of the Singularity made a much bigger and faster splash in science fiction. There, he claimed, it had been anticipated for decades. "More and more, these [science fiction] writers felt an opaque wall across the future. Once, they could put such fantasies millions of years in the future," he wrote. "Now they saw that their most diligent extrapolations resulted in the unknowable . . . soon."[50] Just a few years later, the Singularity was a common feature of science fiction novels, with titles like *Accelerando*, *Singularity Sky*, and *Excession*. "The world we inhabit [is] a world sentenced to Singularity," the science fiction author Ken MacLeod wrote in 2003.[51] By then, the Singularity had become so entrenched within science fiction that, as author and critic Jo Walton put it, "most SF being written now has to call itself 'post-Singularity' and try to write about people who are *by definition* beyond our comprehension, or explain why there hasn't been a Singularity."[52] Author Charles Stross was somewhat less delicate. The Singularity, he said in 2006, is "the turd in the punchbowl of near-future SF. You may politely pretend it isn't there, but everyone has to deal with it."[53]

This was the Singularity coming back to its original home.[54] In 1950, a year before Good first wrote about a "superhuman" machine doing its own programming, *Astounding Science Fiction* published a story by a star of the genre, one Isaac Asimov. The story, titled "The Evitable Conflict," takes place in the year 2052. It describes a meeting between Susan Calvin, an expert in robotics—a word Asimov himself had coined in a story several years earlier—and Stephen Byerley, the coordinator of the world government. But Byerley has less power than his title implies, because the world is effectively run by large "Machines": disembodied, robotic "positronic brains" that dictate the best courses of action for essentially all economic, scientific, and political activity. Despite the supposed infallibility of the Machines, Byerley has noticed strange discrepancies in the world economy. But he knows that the Machines can't be doing this intentionally, because, like all robots, they're programmed to obey the Three Laws of Robotics, the first of which states that no robot can harm a human under any circumstances. So, Byerley says to Calvin, the first thing he did was ask the technicians responsible for the Machines to perform a diagnostic on them, to confirm they're working as they should. To his surprise, he finds that this is impossible, because the Machines have become far too complex for any human to understand: "A team of mathematicians work several years calculating a positronic brain equipped to do certain similar acts of calculation. Using this brain they make further calculations to create a still more complicated brain, which they use again to make one still more complicated and so on. . . . What we call the Machines are the result of ten such steps."[55] The Singularity, avant la lettre.

This wasn't even a new idea in science fiction at the time. Asimov was echoing the 1932 short story "The Last Evolution," by John W. Campbell Jr.[56] Campbell later edited Asimov's robot stories at *Astounding Science Fiction*, and Asimov said he "godfathered the robots."[57] But while Campbell's story did include the same idea of computers designing more advanced computers, Asimov's story also contains a clear articulation of the core motivation behind the Singularity. As the end of "The Evitable Conflict" reveals, despite Byerley's doubts, the Machines are infallible. The solution to his puzzle is merely that the Machines have determined that some humans are trying to thwart their governance, and consequently the Machines have taken steps to keep those people safely out of the way without harming them or disrupting society as a whole. "The Machine is conducting our future for us not only simply in direct answer to our direct questions, but in general answer to the world situation and to human psychology as a whole," Calvin tells Byerley. "The Machine cannot, *must* not, make us unhappy."[58] Utopia, algorithmically guaranteed. In just the same way, the Singularity's superintelligent, benevolent machines will purportedly reduce the multifarious wicked problems of politics, economics, and sociology to a single matter of sufficiently clever computer programming for a sufficiently powerful computer. This is the real appeal behind the modern idea of the Singularity: the seductive promise of all questions, needs, and desires fitted to the single Procrustean solution of code and compute.

* * *

In the first few weeks of 2020, I came across a disturbing graph depicting what was, unquestionably, exponential growth. The graph was attached to a tweet, part of a thread I'd stumbled upon about the new coronavirus, COVID-19. Case rates in the United States were still quite low at the time, with only a handful of cases reported. But this tweet pointed out, although there weren't many cases yet, that was poised to change, fast: left unchecked, cases in the United States would grow to the hundreds of thousands by the end of March. I was alarmed, and immediately started trying to figure out reasons this could be wrong. I couldn't find any obvious problem with the pseudonymous tweeter's claims. But it just seemed so unlikely, on the face of it, that they could be correct. If they were right, everything was about to change, fast and unpredictably, yet there seemed no obvious herald that such a change was coming. How could they be right? Unsure, I went on with my day, troubled by new doubts, trying to figure out if I was worried about something real or had simply been scared by a mirage conjured by a random internet person. Several weeks later, I had my answer.

It's tempting to simply dismiss Kurzweil's claims out of hand. They seem too outlandish, too implausible to take seriously. If such a huge change is coming, there should be signs, something to herald that we are on the verge of an enormous shift, fast and unpredictable, in every facet of human society. How could he be right? But Kurzweil certainly is right about at least one thing: humans have trouble recognizing and accepting the possibility of exponential growth, even when we're already deeply familiar with the concept, as I was in early 2020. "Almost everyone I meet has a linear view of the future," he writes. "That's why people tend to overestimate what can be achieved in the short term (because we tend to leave out necessary details) but underestimate what can be achieved in the long term (because exponential growth is ignored)."[59]

So we can't just laugh off Kurzweil's ideas. He's not crazy, and he's not stupid. Nor is he alone. Today, the Singularity isn't just something that fringe groups like the Extropians take seriously. Kurzweil's books have been bestsellers. His take on the Singularity has landed him in major magazines like *Time*, *Wired*, and *Rolling Stone*. He is regularly invited to speak at major media events like South by Southwest. Singularity University, where "technology experts and entrepreneurs with a passion for solving humanity's grand challenges" can "exchange ideas and facilitate the use of rapidly developing technologies," was cofounded by Kurzweil in 2008 with funding from Google, Autodesk, and other Silicon Valley firms.[60] And in 2012, Kurzweil took up a job at Google as a director of engineering, upon the personal invitation of Google cofounder and then CEO Larry Page.[61] "These are ideas with tremendous currency in Silicon Valley; these are guiding principles, not just amusements, for many of the most influential technologists," explains the computer scientist Jaron Lanier.[62] Many leaders in tech—including Bill Gates and Elon Musk—think highly of Kurzweil and his ideas. "Ray Kurzweil's Moore's Law abstraction is the most important thing ever graphed," says billionaire and tech venture capitalist Steve Jurvetson. "Its continuity—over his lifetime of writing—is the greatest take-away for the future of humanity, and the future of intelligence."[63]

The idea of the Singularity has even made its way into the halls of academia, where the concept has received some attention from philosophers. Bostrom, who was a philosophy professor at Oxford University for over fifteen years, has written extensively about a hypothetical intelligence explosion, both in academic papers and his 2014 book *Superintelligence*, a surprise bestseller heartily endorsed by Bill Gates. "I think at some point we will create machines that are superintelligent," Bostrom said shortly before that book was released. "It's a very mainstream opinion among experts to think that there is a real chance that this may happen over the next few decades, or at least in this century."[64] David Chalmers, professor of philosophy at New York University, wrote a lengthy paper on the subject in 2010. "The singularity idea is clearly an important one," he wrote. "The argument for a singularity is one that we should take seriously."[65] Chalmers says that the idea is based on premises that shouldn't be too hard to swallow. "Where the tech world is concerned, exponential growth claims aren't especially extraordinary. They're quite common," he told me. "[If] we get to the point where AI systems are working at roughly the level and speed of humans, and we've just got some exponential increase in hardware speeds, doubling every five years or whatever you like, then it's just obvious that as a result . . . [y]ou get the intelligence curve to shoot up exponentially."[66] And while Chalmers and Bostrom were among the first philosophers to take the idea of a Singularity seriously, they're not alone anymore: Bostrom was the founding director of the Future of Humanity Institute (FHI) at Oxford, which operated from 2005 until 2024, and where the Singularity was a regular subject of research.

So Kurzweil isn't a lone crank. He's the public face of a movement. And the data that he's assembled to illustrate his law of accelerating returns at work over the course of history are particularly striking. But a closer look at Kurzweil's chart reveals a problem. The inverse of the exponential function is the logarithm. Think of it as a function that tells you roughly how many zeros appear at the end of a number (or, more generally, how many digits the number has). The logarithm of 1,000 is three times larger than the logarithm of 10; the logarithm of a million, 1,000,000, is three times larger than that. Logarithms are more familiar than they seem. We frequently think logarithmically, especially about the past: What were you doing this time last month? A year ago? Ten years ago?

Chuck Klosterman, the author and critic, illustrates this point well with a simple game: without looking anything up, name a definitely real person who lived in each century, working backward from the twenty-first. For this century and the last, it's easy to think of many names, including people we personally know. The nineteenth century isn't too difficult either, though you're probably going to think of famous political or historical figures (Harriet Tubman, Napoleon). The eighteenth century is a little harder, but not too difficult (Benjamin Franklin). Things get harder still around the seventeenth century, and you're probably down to just a few names that you're sure about, if any (Isaac Newton). And unless you have specialized domain knowledge about particular areas of history, it's not likely that you can get much past the sixteenth century (Shakespeare, Galileo) or fifteenth century (Joan of Arc). As Klosterman puts it, if someone makes it to the twelfth century, "it usually means they either know a lot about explorers or a shitload about popes."[67]

But this doesn't mean that you can't name anyone who lived earlier than around 1400. Your knowledge just gets spottier. Mansa Musa, Eleanor of Aquitaine, and Ibn Rushd are in there somewhere; further back, there's religious and philosophical figures like Muhammad, Jesus, and Confucius; and then there's a smattering of political leaders like Cleopatra, King Tut, and Hammurabi—and if you're up on your internet memes about cuneiform complaint letters, then way back near the start of recorded history there's Ea-Nāṣir and Ibbi-Ilabrat.[68] But you can't go linearly through history anymore. Your historical knowledge has turned logarithmic. This isn't because everything was boring in, say, the seventh century—the myth of the Dark Ages is just that, a myth, and also there are plenty of other places in the world outside of Europe where things were happening at the time—it's just a matter of your perspective from this particular point in history. Three hundred years from now, if the United States still exists, it's quite likely that the average American wouldn't be able to name a single president from the twentieth century at all, just as we can't reliably name political leaders from the 1600s. Memories fade, and the selection of salient events that we remember out of history is, at best, only loosely tied to their importance. We have a logarithmic view of history because we can't possibly learn and retain everything that ever happened before we were born.

Seen through this lens, Kurzweil's trends become more suspect. It seems likely that he's confusing a logarithmic view of history for an exponential trend in biological and technological development. His list of biological milestones gives this away: rather than picking particularly important events in the evolution of all life on Earth, he's mostly chosen milestones leading up to the evolution of *humans*, as if humans are the ultimate goal of evolution. (Evolution has no goal, as Kurzweil surely knows.) This kind of cherry-picking makes it easy to create the appearance of an exponential trend. The list of scientific and technological advances is also cherry-picked: among the many missing milestones are the invention of gunpowder in China in the ninth century, the development of Newtonian physics in the seventeenth century, the germ theory of disease, the periodic table of the elements, the railroad, and the telegraph. None of these fit the exponential pattern, and all of them were easy to find, because there are many time periods that have no representation whatsoever on Kurzweil's list.* These omissions

* Going through the many flaws in Kurzweil's exponential trend line would take an entire chapter on its own. I will restrain myself here to noting three more:

1. Kurzweil is helped by the fact that there's just less evidence of things deeper in the past,

are unlikely to be deliberate, any more than the near total exclusion of technological or cultural developments outside of Eurasia. Both are just the result of a biased perspective limited by a particular place and time. It's like looking out from the top of a skyscraper: nearby, everything looks big and important; off in the distance, only a few salient points are visible. But things aren't actually getting bigger as they get closer to you, nor are you situated in any particularly important or unique location. Kurzweil is inferring a kind of teleology, a purpose and trend, where none exists.

Or almost none. It's true that some exponential trends exist in the history of technology, like Moore's law. But on closer examination, the fate of Moore's law itself provides further evidence that Kurzweil is wrong. "It actually depends on your definition of Moore's law," says Tsu-Jae King Liu, dean of the College of Engineering at UC Berkeley and member of the board of directors of Intel. Moore originally pointed out two different exponential trends: an exponential increase

---

and so we simply know less about periods further back than we do about more recent times. For example, the rarity of truly ancient fossils means it's harder to know about major evolutionary milestones the further back we look. Indeed, it's actually possible that Kurzweil is wildly incorrect in a surprising way: if there was a species that lived one hundred million years ago on Earth that had a technological civilization comparable to our own, it is entirely possible that we would be unable to detect that their civilization had ever existed. See Jason T. Wright, arXiv:1704.07263; Gavin A. Schmidt and Adam Frank, arXiv:1804.03748.

2. Some time periods are represented on Kurzweil's list by a single point, like the Cambrian explosion or the Industrial Revolution. If he had instead included individual events during those periods, the exponential trend would be destroyed, or at least much noisier.

3. The last four entries in my list of technological and scientific omissions are all from the nineteenth century. There is a compelling argument, due to Cosma Shalizi, that the Singularity is "near" in that it *already happened* in the 1800s. "Why, then, since the Singularity is so plainly, even intrusively, visible in our past, does science fiction persist in placing a pale mirage of it in our future? Perhaps: the owl of Minerva flies at dusk; and we are in the late afternoon, fitfully dreaming of the half-glimpsed events of the day, waiting for the stars to come out." Cosma Shalizi, "The Singularity in Our Past Light-Cone," http://bactra.org/weblog/699.html.

in the number of transistors per integrated circuit, and an exponential decrease in the cost per transistor. Kurzweil makes use of both forecasts in *The Singularity Is Near*, extending them far into the future. But "the cost per transistor actually has gone up in the most recent generations. . . . [So] if it's strictly lower cost per transistor, then yes, [Moore's law] has stopped," Liu says. "If it's that the number of transistors on the most advanced integrated system is roughly doubling every two or so years, then Moore's law is actually still continuing. . . . The total number of transistors integrated into a package has indeed continued to increase at an exponential pace."[69]

Liu herself has helped keep this form of Moore's law alive: she is the coinventor of the FinFET transistor, currently the state-of-the-art transistor design. But, she says, the survival of this version of Moore's law is due to the particular phrasing of his prediction, involving the number of transistors placed onto a single chip. This is still increasing exponentially, and will do so through the end of this decade, says Liu, because transistors are now being packed on top of each other vertically. "The number of transistors per unit *volume* will continue to grow (but not at an exponential pace) as layers of transistors are stacked over each other and as chips are stacked over each other with increasingly higher density," Liu tells me.[70] The transistors themselves are still shrinking, too, but that can't continue indefinitely. "The rate at which transistor lateral dimensions continue to scale down with each new generation of chip manufacturing technology will slow down or even stop, due to physical limits and/or financial considerations."[71] The physical limits Liu is alluding to arise somewhere around 0.1 to 1 nanometer, because atoms of silicon themselves are about 0.2 nanometers across, and you can't build a transistor out of less than a single atom. This limit isn't possible to break, and we're on the verge of reaching it—the smallest transistors yet made are indeed around 0.3 nanometers across. Silicon transistors aren't going to get appreciably smaller, ever. The exponential trend is over. (Liu did say the exponential trend in transistors packed onto a single chip would continue, but that has to end too, and sooner rather than later. Since the volume of a single transistor can't shrink much more, that means that either the number of transistors on a chip won't keep increasing exponentially for long, or the chips will grow in height exponentially—and it's not going to be the latter.) Gordon Moore himself saw this coming. "It can't continue forever. The nature of exponentials is that you push them out and eventually disaster happens," he said in an interview in 2010. "In terms of size [of transistor] you can see that we're approaching the size of atoms which is a fundamental barrier. . . . We have another 10 to 20 years before we reach a fundamental limit."[72]

But the problems for Kurzweil's law of accelerating returns aren't just due to physical limits. To get steady growth like Moore's law, "you can't just [say], 'OK, computing algorithms are going to run faster and so we can innovate the next generation of technology faster,'" Liu tells me. "You actually still need some innovation, some pioneering creative ideas to sustain this so-called Kurzweil's law. I think maybe Kurzweil's law is a little too simplistic, because it doesn't take into account the fact that complexity also increases exponentially with advancements in technology."[73] The history of Moore's law bears this out. From 1971 to 2014, the heyday of Moore's law, the number of transistors crammed onto a chip increased by a factor of about two million. This was steady exponential growth, doubling just about every two years. But maintaining that steady growth required exponential growth in effort and resources too: in 2020, a team of economic researchers at Stanford and MIT found that, over that same time period, the investment in semiconductor R&D funding and personnel increased eighteen-fold just to maintain that constant growth rate. "Put differently," they wrote, "it is *around 18 times harder* today to generate the exponential growth behind Moore's Law than it was in 1971." Their conclusion is that research productivity has declined because "ideas, and the exponential growth they imply, are getting harder to find."[74] Moore's law, then, isn't an example of accelerating returns—it's an example of *diminishing* returns for the same level of investment.

The fate of Moore's law is the fate of all exponential trends: they end, just as Moore himself said. Either things slowly get harder, flattening out the growth curve from exponential to S-shaped (known as a logistic curve; see Figure 2.2), or they reach a firm limit and just stop, often followed by a crash. Put a few bacteria into a plate of nutrients, and their numbers will grow exponentially—until they run out of food, at which point growth halts and they all die. In *The Singularity Is Near*, Kurzweil illustrates exponential growth by talking about lily pads growing exponentially in a pond, doubling the amount of water they cover every few days. If they cover 1 percent of the pond on the first day, he says, then a few weeks later the pond will be completely covered.[75] That's true, but that's also where the lily pads' growth ends, because they can't cover more than 100 percent of the pond. Every exponential trend works like this.[76] All resources are finite; nothing lasts forever; everything has limits. This is the crucial flaw at the heart of Kurzweil's argument. Exponential trends exist—but the only thing absolutely guaranteed about the future of an exponential trend is that, sooner or later, it will end. Returns diminish. Extrapolating exponential trends indefinitely into the future is to confidently assert exactly what won't happen.

This isn't news. In 1960, a group of researchers at the University of Illinois noticed that human population growth over the previous two thousand years wasn't just exponential but hyperbolic—growing faster than any exponential curve. Extrapolating the curve, they found a surprising result: on November 13, 2026, the population of humanity would become infinite. "Our great-great-grandchildren will not starve to death. They will be squeezed to death," they concluded with tongue planted in cheek, making it quite clear they knew that the hyperbolic trend in human population growth would end soon.[77] And they were right: by 1964, the annual growth rate of the world population was slowing. It was 2.2 percent at its peak in 1963, 2.0 percent in 1973, and as of 2023 a modest 0.9 percent. The United Nations currently forecasts that population growth will halt around 2080, give or take a couple of decades, and then the population will slowly decline from a maximum of about ten billion people, not much more than the eight billion who are alive today. The projected curve of the human population over time fits a near-perfect logistic S-curve, meeting the same fate as most exponential growth in the real world.

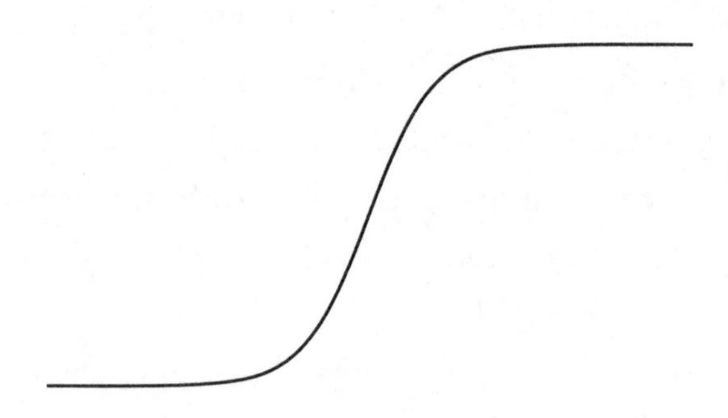

Figure 2.2: The S-curve of logistic growth.

To his credit, Kurzweil knows that exponential growth always turns into logistic growth. He even agrees that Moore's law can't continue in silicon due to immutable physical limits. But this doesn't dissuade him. "S-curves are typical of biological growth," he writes. "[But] the overall exponential growth of an evolutionary process (whether molecular, biological, cultural, or technological) supersedes the limits to growth seen in any particular paradigm (a specific S-curve) as a result of the increasing power and efficiency developed in each successive paradigm. The exponential growth of an evolutionary process, therefore, spans multiple S-curves."[78] To illustrate this, Kurzweil points to the history of computer technology since the turn of the twentieth century. First, he says, there were electromechanical computers; after those came vacuum tube computers; after that, discrete transistors; and finally we came to integrated circuits formed on silicon wafers, which is still the technology that computers run on today.[79] During each of those eras, says Kurzweil, the power of computing machines did indeed describe an S-curve: exponential growth that eventually hit a ceiling as the limitations of that particular technology were reached. But, he points out, those S-curves combine to keep the overall exponential trend in computational power going. (See Figure 2.3.) "A specific paradigm . . . generates exponential growth until its potential is exhausted. When this happens, a paradigm shift occurs, which enables exponential growth to continue."[80]

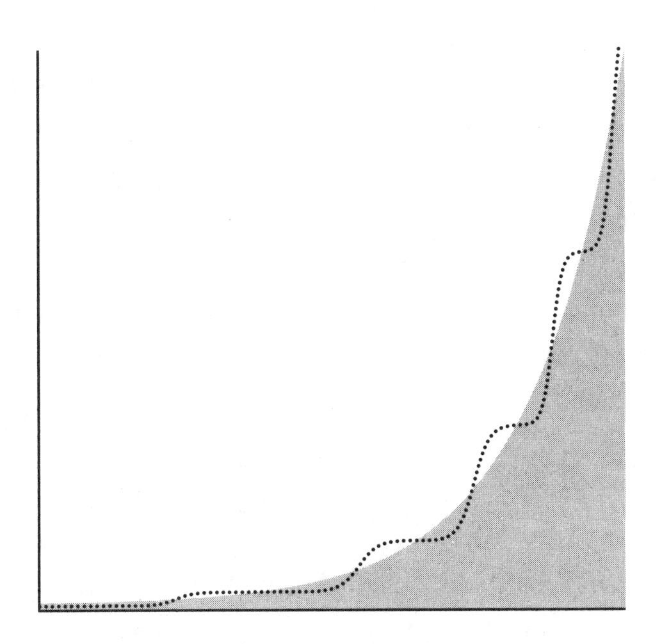

Figure 2.3: Exponential growth from stacked S-curves.

This is just a combination of more cherry-picking with flat-out wishcasting. Kurzweil's stacked S-curves ignore the millennia of mechanical calculation technologies before the twentieth century, from the abacus to the Antikythera mechanism to Charles Babbage's analytical engine. Those don't follow anything like the trend he's describing. And even putting that earlier history aside, there's no guarantee that new computing technologies will be developed on the exacting schedule that Kurzweil's trend demands. New technologies might get harder and harder to develop; a new technology might be developed that looks promising but has unavoidable problems that prevent it from being implemented; or other factors might get in the way of a new technology, like economic pressures, societal changes, or scarcity of necessary resources.

This has happened over and over again in the history of human technology. Exponential trends based on technological abilities peter out. Two hundred years ago, the fastest a human being could reliably go was around the top speed of a horse, roughly 35 mph (55 kph). This had been true for thousands of years. But then, steam trains were developed, matching the top speed of a horse by 1830 and then surpassing it. After steam trains came internal-combustion-powered cars. After cars came airplanes; after airplanes, liquid-fueled rockets. The top speed a human being had ever achieved rose exponentially over the course of the 1800s and well into the 1900s. But on May 26, 1969, that stopped. That day, Thomas Stafford, Eugene Cernan, and John Young traveled at 24,816 mph (39,938 kph) as they returned to Earth after orbiting the Moon on *Apollo 10*. No human has gone faster since. Even if we were to send humans to Mars, they wouldn't hit a top speed much faster than the crew of *Apollo 10*, if at all. And if we expand our definitions to include space probes and other artificial objects, that still doesn't help much. The record for fastest thing made by humans is currently held by the Parker Solar Probe, which (as of this writing) is scheduled to streak past the Sun at over 430,000 mph (690,000 kph)—0.064 percent of the speed of light—by the end of 2024. This almost triples the previous record, set by the *Helios 2* probe in 1976 as it made its closest approach to the Sun. It's not a coincidence that both of these ludicrously speedy machines were traveling close to the Sun—that's precisely why these are the voyages at the top of the speed record list. The Sun is not just at the center of the solar system; it's at the bottom of it, gravitationally speaking. The Sun contains 99.86 percent of all the mass in the solar system, with a mass over 300,000 times that of Earth. Solar probes can hit such high speeds because they're falling toward that enormous mass from a height of nearly a hundred million miles (150 million kilometers) without any air resistance to slow them down. After millennia of technological advances, humanity's best method for making things go really fast is to drop them.

Why can't we make things go faster? The historical reasons for this are complicated; arguably, the root cause is political. But the only reason political budgets are germane to the conversation is that the resources required to go appreciably faster than *Apollo 10* or the Parker Solar Probe are so phenomenally huge that only large nations (and, perhaps, multinational corporations) can muster them. So while the details of the historical reasons are complex, the root cause for the S-curve is fairly simple: there was no breakthrough technology developed that made it easier to go faster, because it's very difficult to go much faster than we already have.

Diminishing technological returns aren't limited to computer chip design and traveling faster. The 2020 Stanford-MIT research on Moore's law was part of a broader study, which came to a similar conclusion about the entire economy. The authors of that study provide compelling evidence that research productivity today—imperfectly measured as the translation of industrial R&D into economic growth—is over forty times lower than it was in the 1930s. Growth has been sustained despite these diminishing returns because of an exponential increase in resources and careers devoted to research over the same time period. That trend can't last. As for why research productivity has declined so dramatically, the authors' overall conclusion is the same as it was for the semiconductor industry: "Ideas are getting harder and harder to find," they wrote. "Put differently, just to sustain constant growth in GDP per person, the United States must double the amount of research effort every 13 years to offset the increased difficulty of finding new ideas."[81] This makes some intuitive sense. In any new field, there is low-hanging fruit; once that is picked, finding new good ideas becomes harder. And as the many scientific revolutions of the nineteenth and early twentieth centuries continue to age, the low-hanging technological fruit they enabled has become increasingly picked over. There are new ideas. There will probably always be new ideas. But finding them, and finding ways to use them, is getting harder.

New computing technologies are no exception to this trend. Integrated circuits are near the end of the line, and there's no obvious replacement in sight that will enable Kurzweil's version of Moore's law to continue on much past the end of this decade.[82] In 2005, Kurzweil predicted that "three-dimensional molecular computing" using carbon nanotubes would be the next paradigm in electronics, taking over when silicon chips exhausted their potential—right around now. That technology hasn't panned out yet and doesn't look likely to do so soon, if ever. Today, the fundamentally new computing technology garnering the most attention is quantum computing. But quantum computing isn't touted as a replacement for existing computers and can't continue Moore's law. Instead, the promise of quantum computing lies in its theoretical ability to perform a few specialized tasks dramatically more efficiently and quickly than conventional computers, such as cryptography, molecular modeling, and solving certain kinds of optimization problems. Quantum computers might be able to perform many impressive feats, but even if they fulfill their promise, they likely won't lead to faster computers overall.

Yet even if some new technology does save Moore's law for a while, that doesn't mean the Singularity is back on schedule. If computers continue to grow in power exponentially, that doesn't imply that an AGI running on a computer will grow in intelligence exponentially too. Intelligence doesn't necessarily correspond to computational power in a simple way. "If we do want a notion of intelligence that scales linearly with compute power, it's not going to be the notion of intelligence that grounds the kind of things we want," says David Thorstad, a philosopher at Vanderbilt University. "So if it's also supposed to scale with the ability to predict and act on the world, or play games, or pass intelligence tests, or do anything we would standardly define as intelligence, it's not going to scale linearly with compute power."[83] In fact, throwing more computational power at solving specific problems often yields—yet again—diminishing returns. A study done in 2022 by researchers at MIT and the University of Brasilia showed that exponential increases in computing power led to merely linear increases in computers' abilities to predict the weather, solve protein-folding problems, find new sources of oil based on geological data, and play games of Go and chess.[84]

Intelligence is more than problem-solving. But it's not clear what intelligence really is, and even less clear how to measure it. Psychologists have studied intelligence systematically for well over a century; the actual existence of some measure of overall intelligence, which they dub "general intelligence" or just *g* for short, is debatable at best. Statistically, there's no good case for the actual existence of *g*. "The myth of *g* legitimates a vast enterprise of intelligence testing and theorizing," writes statistician Cosma Shalizi, who calls *g* a "statistical myth." "The sooner we stop paying attention to *g*, the sooner we can devote our energies to understanding the mind."[85]

Without a concrete notion of general intelligence, it's not clear what would constitute an "intelligence explosion," nor what it would mean for one computer to be twice as intelligent as another.[86] Defining AGI in the first place looks shaky too. But even if we accept a vague provisional definition for AGI in the spirit of Kurzweil's—something like "an artificial machine that can do everything a typical human adult can do"—it's still not obvious that a modern computer would be capable of running an AGI. In *The Singularity Is Near*, Kurzweil talks a great deal about reverse engineering the human brain: Chapter 4 of the book is titled "Achieving the Software of Human Intelligence: How to Reverse Engineer the Human Brain."[87] Kurzweil estimates that it would take a computer performing 10^{14} to 10^{16} computations per second "to achieve functional equivalence of all brain regions," and that the memory capacity of the human brain is around 10^{13} bits, or roughly one terabyte.[88] "Supercomputers will match human brain capability by the end of this decade and personal computing will achieve it by around 2020—or possibly sooner," he wrote in 2005.[89]

But, he added, "The hardware computational capacity is necessary but not sufficient. Understanding the organization and content of these resources—the software of intelligence—is even more critical." To do that, he claimed, we'd need very detailed scans of the brain—and Kurzweil was quite sanguine about the prospects for such brain scanning. "Our ability to reverse engineer the brain—to see inside, model it, and simulate its regions—is growing exponentially," he wrote.[90] "The resolution of noninvasive brain-scanning devices is doubling about every twelve months. . . . In the 2020s we will be able to observe all of the relevant features of neural performance with very high resolution from inside the brain itself."[91] None of these forecasts have panned out. In fact, the maximum resolution of noninvasive living human brain scans in 2024 is not appreciably better than it was in 2000, providing another example of the S-curve of logistic growth and its diminishing returns, rather than Kurzweil's claimed exponential growth. We still don't have the ability to see what's happening in the living brain at the level of individual synapses or even hundreds of synapses. Such technology might be developed someday, but there's no indication that its invention is coming anytime soon. Kurzweil knows this: he said in 2024 that improvements to brain-computer interfaces haven't kept to his schedule because of safety concerns. Testing things out on real human brains naturally makes people worry about risks. But Kurzweil doesn't see this as a serious problem, because he claims computers may soon let us simulate the brain well enough to build brain scanning and interface devices without testing them on living humans.[92] What he didn't explain was how such simulations could be built without having the brain scanning technology in the first place.

Even if that were possible, existing computers aren't powerful enough to allow for such a plan. Personal computers generally are not much more powerful than a trillion operations per second; that's a hundred times lower than Kurzweil's lowest estimate for the computational power of the human brain. The fastest supercomputers in existence today are capable of 10^{18} computations per second, more than Kurzweil's estimated range for "functional equivalence of all brain regions." But they're still shy of his true upper bound for simulating the human brain, which he estimates as 10^{19} computations per second.[93] And that estimate is itself suspect. It's based on the number of synapses in the human brain—connections between neurons—but those neurons are also performing important functions within themselves, using the phenomenally complex protein structures that form the basis for all life. Nor do synapses perform anything like the simple on-off binary calculations of a digital computer; they are analog, taking values along a continuum of possibilities. Furthermore, the connections between neurons are mediated by neurotransmitters, which have a tremendous effect on the operations of the brain (indeed, that's how most psychiatric medication works). And then there are the glial cells, which are at least as numerous in the brain as neurons, and which play their own important role in the brain's operation.[94] It's entirely likely that Kurzweil's upper bounds on the brain's computational power and memory capacity are too low by many orders of magnitude. Indeed, a 2008 study coauthored by Nick Bostrom suggested that the computational power of the brain could easily be ten million times greater than Kurzweil's upper bound.[95] "There's a lot of unknown unknowns and known unknowns about what information you actually need out of a brain to make something that would be enough like a person" to simulate a human brain on a computer, says Michael Hendricks, a neurobiologist at McGill University. "Synapses vary in all kinds of different ways from each other, neurons can be in all kinds of different states, depending on ion channel composition and what genes are expressing. They're not just generic widgets, they're different things. And so how much information you need is unknown, but it's probably an amount of information that is more than all the computing power our society will produce for many decades."[96]

If the brain is that powerful, and Moore's law really has come to an end, then computers as we know them now are just not suitable for emulating human-level intelligence. A fundamentally different approach would be needed to build an AGI. More generally, the brain might not be best thought of as a computer at all. "Although I do firmly believe that the brain is a machine, whether this machine is a computer is another question," wrote MIT roboticist Rodney Brooks in 2008. He continued:

> In centuries past the brain was considered a hydrodynamic machine. René Descartes could not believe that flowing liquids could produce thought, so he came up with a mind-body dualism, insisting that mental phenomena were nonphysical.

> When I was a child, the prevailing view was that the brain was a kind of telephone-switching network. When I was a teenager, it became an electronic computer, and later, a massively parallel digital computer. A few years ago someone asked me at a talk I was giving, "Isn't the brain just like the World Wide Web?" We use these metaphors as the basis for our philosophical thinking and even let them pervade our understanding of what the brain truly does. None of our past metaphors for the brain has stood the test of time, and there is no reason to expect that the equivalence of current digital computing and the brain will survive.[97]

The architecture of modern computers, involving a central processing unit connected to working memory and long-term storage, does not resemble our best understanding of the logical organization of the human brain. Indeed, the best science suggests that the human brain isn't structured like a human-made object at all, which makes reverse engineering it extraordinarily difficult.[98] "Brains, unlike any machine, have not been designed. They are organs that have evolved for over five hundred million years, so there is little or no reason to expect they truly function like the machines we create," writes zoologist and historian of science Matthew Cobb. "The brain is not a computer, but it is more like a computer than it is like a clock. . . . Over the centuries, each layer of technological metaphor has added something to our understanding, enabling us to carry out new experiments and reinterpret old findings. But by holding tightly to metaphors, we end up limiting what and how we can think."[99] Modern AI systems based on "neural networks" claim to have an architecture inspired by neurons, but they bear the same resemblance to a real brain that a raincloud emoji has to a nor'easter. Brains and computers just don't operate in the same way at all, and thus concepts taken from computer science aren't generally applicable to the brain. "Even something as apparently straightforward as working out the storage capacity of the human brain falls apart when it is attempted," writes Cobb. "Neurons are not digital. . . . Each brain is continually changing the number and strength of its synapses, and above all it does not work by synapses alone. Neuromodulators and neurohormones are also responsible for the way that the brain functions."[100] And that's just in the brain. As Brooks, Cobb, and others have pointed out, understanding human cognition requires looking beyond the brain. "The physiological reality of all brains is that they interact with the body and the external environment from the moment they begin to develop," writes Cobb. "Animals are not robots piloted by brains, we are all, whether maggots or humans, individuals with agency and a developmental and evolutionary history."[101] This is one of the core lessons of the relatively new field of "embodied cognition": We are not just our brains. We are our bodies, situated in our environment.

If something with the power of the human brain—along with some equivalent to the brain's connections to the rest of the body and the wider world—is required for human-level intelligence and consciousness, then AGI is a long, long way off. "There is no reason, beyond our ignorance of how consciousness works, to suppose that [the Singularity] will happen in the near future," writes Cobb. "The scale of complexity of even the simplest of brains dwarfs any machine we can currently envisage. For decades—centuries—to come, the singularity will be the stuff of science fiction, not science."[102] There are theoretical arguments that suggest an AGI should be possible to build with the kinds of computers we currently use, if they're powerful enough.[103] But even if those arguments are correct, and even if a powerful enough computer could be built, it might be that building an AGI with existing computers is possible in theory but prohibitively difficult in practice—especially when we have neither a firm theoretical grasp of how an AGI would work in the first place nor a detailed theory of how the brain gives rise to consciousness. We don't even have a good definition of what AGI is.

A different kind of computer, or a different kind of machine altogether, might make AI research much easier.[104] That would be welcome news, since work toward AI has almost always been harder than computer scientists and cognitive scientists guessed it would be. "Every path to AI has proved surprisingly difficult to date. The history of AI involves a long series of optimistic predictions by those who pioneer a method, followed by periods of disappointment and reassessment," writes David Chalmers.[105] This has been true in the field since the start. In the summer of 1956, an interdisciplinary team of researchers met for eight weeks to discuss a new field that the organizer, computer scientist John McCarthy, called "artificial intelligence." "We propose that a 2 month, 10 man study of artificial intelligence be carried out during the summer of 1956 at Dartmouth College in Hanover, New Hampshire," McCarthy and his co-organizers wrote in a grant proposal for the workshop, submitted the year before. "An attempt will be made to find how to make machines use language, form abstractions and concepts, solve kinds of problems now reserved for humans, and improve themselves. We think that a significant advance can be made in one or more of these problems if a carefully selected group of scientists work on it together for a summer." Seventy years later, we still don't have any computer program that can do all the things on their list. There have been AI booms—like the one that exploded into public consciousness in 2022, powered by large language models (LLMs)—and, to date, they have all been followed by "AI winters," when progress slows and the state of the art stagnates for years or decades until the next breakthrough. And Brooks thinks winter is coming. "[LLMs] are following a well worn hype cycle that we have seen again, and again, during the 60+ year history of AI," he wrote in 2024. "Get your thick coats now. There may be yet another AI winter, and perhaps even a full scale tech winter, just around the corner. And it is going to be cold."[106]

All of this uncertainty reflects the fact that we just don't know what it will take to build a machine that can do all the things a human can do. In the face of that ignorance, Kurzweil's claims of exponential trends in the future of technology in general, and AGI in particular, seem excessively confident. As such, they face the same burden of proof as any other extraordinary claim: they require extraordinary evidence. That evidence is lacking. Kurzweil isn't crazy, and he isn't stupid. He's just mistaken. But now, twenty years after he first predicted a Singularity by 2045, after so many failed predictions of new technologies along the way—everything from cellular-level brain scanners to RNA-based weight-loss drugs, from the rise of carbon-nanotube transistors to the continued reign of Moore's law—Kurzweil still thinks the Singularity is coming, soon. In March 2024, he said that he thinks the world is still on track to hit the Singularity in 2045. Later that year, his new book, titled *The Singularity Is Nearer*, was published aloft another wave of hype. And unlike sci-fi writer Vinge, Kurzweil thinks that he can pierce the veil of magic-seeming tech on the other side of the Singularity. "Despite our profound limitations of thought, we do have sufficient powers of abstraction to make meaningful statements about the nature of life after the Singularity," he wrote in *The Singularity Is Near*.[107] In that future, he says, we'll live in a world devoid of danger, pain, and hardship—and just about everything else too.

* * *

Nearly twenty years before *The Singularity Is Near*, K. Eric Drexler published his own blueprint for the future. Drexler, a researcher at MIT who had arrived there as a teenager over a decade earlier with dreams of space, had turned his intellect and imagination toward the world of the very small.[108] There, he claimed, lay the escape from the limitations of Earth that he'd sought. Inspired by the intricate and complex procession of proteins that operate in living cells, Drexler envisioned a future of nanotechnological wonders in his 1986 book *Engines of Creation*. The first step, he said, would be to construct organic machinery similar to the ribosomes and RNA that translate DNA's genetic code. That first generation of nanomachines would then be used to construct a second generation of machines, made of "tougher stuff" like diamond that didn't require water to function—and that would allow for "control over the structure of matter" down to the atomic level. "Molecular machines of the second generation will be able to build virtually anything that can be designed," Drexler claimed.[109] Crucially, that included more molecular machines of the same type: one machine could make another, and then each could build one more, then those four could duplicate themselves, and so on, with the inexorable mathematics of exponential growth leading to a rapid explosion in the number of nanomachines available to build whatever humans could dream of.

The rise of these nanomachines—or "molecular assemblers," as Drexler dubbed them—would, he said, bring about a total revolution, not only in manufacturing, but in all aspects of life and human civilization. Assemblers, Drexler claimed, would create permanent abundance of energy and food; they would eliminate illness and cure "a disease called aging"; they would enable the development of advanced AI. And they would dramatically speed up and miniaturize all manufacturing, while allowing literally anything to be manufactured.[110] Nanotech, according to Drexler, would be faster, cheaper, and more powerful than any other technology in the history of humanity. In *Nanosystems*, Drexler's 1992 follow-up to *Engines of Creation*, he claimed that nanotech engines could output a trillion kilowatts per cubic meter. "It is difficult . . . to get one's head around the utter raw power and potential of real nanotech," wrote J. Storrs Hall, an engineer, author, and colleague of Drexler's, in 2021.[111] To illustrate this, Hall made an estimate, based on Drexler's work, of how long it would take to reconstruct all the "capital stock" of the United States—"to rebuild every single building, factory, highway, railroad, bridge, airplane, train, automobile, truck, and ship," as Hall put it—starting with a single kilogram of "mature nanotechnology." Hall's answer, relying on the exponential growth of Drexler-style nanomachines, was one week.[112]

Drexler wasn't the first person to think about the possibility of such small machines. The physicist Richard Feynman gave a talk on the idea in 1959, and a few years later Marvin Minsky wrote about it in one of the essays he contributed to *The Scientist Speculates*. But Drexler was the first to give a detailed picture of what such nanobots might mean for expanding the capabilities of humanity. His vision of a limitless technology fueled a profusion of dreams and desires. Hall claimed that nanotech would provide "immortality in peak physical health" and power flying cars, along with a myriad of other retro-futuristic fantasies straight out of *The Jetsons*.[113] Drexler wrote in *Engines of Creation* that nanotech would make cryonics—the cryogenic freezing and revival of the recently dead—a viable technology; he even signed up for cryonic preservation in the event of his own death. Drexler also wrote that nanotech would make space travel far simpler, faster, and more comfortable. Unsurprisingly, the nascent transhumanist movement and other futurists in the 1980s jumped on Drexler's ideas. The Extropians frequently talked about nanotech on their email list in the '90s, with Drexler himself sometimes contributing to the discussion online and occasionally in person at Extropian conferences. "I agree with most of the Extropian ideas," Drexler said. "Overall, it's a forward-looking, adventurous group that is thinking about important issues of technology and human life and trying to be ethical about it. That's a good thing, and shockingly rare."[114]

In Drexler's work, Kurzweil saw the means to reshape not only the Earth but the entire universe. "Nanotechnology promises the tools to rebuild the physical world—our bodies and brains included—molecular fragment by molecular fragment, potentially atom by atom," he writes.[115] Atoms are around two-tenths of a nanometer across, on average; that's eight billionths of an inch and one five-millionth of a millimeter, half a million times smaller than the thickness of a piece of paper. There are 10^{24} atoms in a heaping spoonful of sugar; there are around 10^{50} on and in the Earth itself. Across the universe, there are roughly 10^{80} atoms in total, a one with eighty zeros after it.

Kurzweil wants to use all, or nearly all, of these atoms for building computers. "The law of accelerating returns will continue until nonbiological intelligence comes close to 'saturating' the matter and energy in our vicinity of the universe with our human-machine intelligence," he writes. "By saturating, I mean utilizing the matter and energy patterns for computation to an optimal degree, based on our understanding of the physics of computation."[116] He envisions nanotechnology on a galactic scale, with fleets of nanomachines rewriting the structure of alien solar systems, filling their stars and planets with circuitry. "We could send swarms of many trillions of them, with some of these 'seeds' taking root in another planetary system and then replicating by finding the appropriate materials, such as carbon and other needed elements, and building copies of themselves," writes Kurzweil.[117] Each "nanobot colony" would in turn produce more fleets of nanobots, until the entire accessible portion of the universe is filled, turning the cosmos into a single enormous computer. Once that's accomplished, Kurzweil estimates the resulting universal supercomputer would be able to perform about 10^{90} calculations per second and store a comparable number of bits.[118] All that computing power and storage won't just be for our AIs, says Kurzweil; it'll be for us, too, once we merge with our machines. The machines will get exponentially more powerful as our biological brains remain the same, meaning an increasing amount of our thinking will happen on machines until we're virtually entirely machine ourselves, living inside the computer that has replaced the universe.[119] That, or the nanobots will simply scan our brains and upload the details of our brain states into the computer, allowing copies of ourselves to live purely digital lives.[120]

If we merge with our machines or become digital uploads, then the fleets of nanomachines building computer circuitry with all available atoms would be turning all the stars and planets, all the galaxies and nebulae, every rock and mote of dust around a distant star, into pieces of our new synthetic brains. This, Kurzweil says, is when "the universe wakes up": "The 'dumb' matter and mechanisms of the universe will be transformed into exquisitely sublime forms of intelligence," he writes.[121] To Kurzweil, this is inevitable, a universal process that would happen to any civilization, here or elsewhere in space. "Once a planet yields a technology-creating species and that species creates computation (as has happened here), it is only a matter of a few centuries before its intelligence saturates the matter and energy in its vicinity, and it begins to expand outward at at least the speed of light," he writes. "Such a civilization will then overcome gravity (through exquisite and vast technology) and other cosmological forces—or, to be fully accurate, it will maneuver and control these forces—and engineer the universe it wants. This is the goal of the Singularity."[122]

Kurzweil claims this goal is so universal that he derives another surprising conclusion from it. "We are probably alone in the universe. . . . Based on the law of accelerating returns, once an ETI [extraterrestrial intelligence] reaches primitive mechanical technologies, it is only a few centuries before it reaches the vast capabilities I've projected for the twenty-second century here on Earth." And if an alien civilization had already done that, he writes, "it would be hard to miss."[123]

Kurzweil's claim of inevitability is strikingly strident, since his universal plan depends on wholly unproven technology. All of it—from the detailed brain scanning to the computers powerful enough to emulate a brain to the nanotechnology itself—is highly speculative, with no extant proof of concept or plausible schematics in existence. "What you actually have to do in a computer to make something that would be sentient, no one knows what that is. No one knows what scale of detail or granularity you need in a simulation," Hendricks tells me. "And you can't just simulate the brain. You have to simulate everything about the body, what the embodiment of being in the world is. You have to simulate everything about a world. None of it's trivial. And [transhumanists] would say, 'Of course, it's not trivial.' They'll just say, 'There's nothing that says we can't do it if we try hard enough.' And that's not how the world works!" And, he adds, getting close to success might be worse. "The trial-and-error risks here are pretty awful. Let's say we start to get close to making a sentient representation of a human brain in a computer. . . . If you have a small difference in the information your eye is giving your brain and your ear is giving your brain, that's already an awful feeling. It's like seasickness, and nausea, or different types of pain. So what we're promising to do here is to create thousands or millions of instances of sentient beings in computers that are probably suffering horribly, and are just going to get turned off. I mean, you could see this really macabre process of creating—if you imagine you can—sentient things in computers. There's a lot of things to get wrong. And those outcomes are terrible."[124]

There's an even deeper issue here, though. Put the hellish torture of an improperly uploaded brain aside, along with the thorny philosophical questions about whether a digital copy would actually be you. Even put aside questions about how exactly traveling light-years from star to star or galaxy to galaxy would work without technology straight off of the starship *Enterprise*. There's still a serious problem with Kurzweil's notion of waking up the universe: it's a euphemism for total destruction. It would be the end of nature, colonialism on a universal scale, with entire galaxies' worth of planets and stars chewed up to provide more computing power for the digital remnants of humanity. Hence Kurzweil's insistence that alien life is unlikely: it is an assurance that the universe is ours for the taking, with nobody else there to worry about.

This line of reasoning is morally abhorrent, not to mention scientifically specious. There is no good way of reliably estimating the prevalence of alien life with the level of specificity that Kurzweil is claiming. (And it seems rather convenient that Kurzweil is optimistic about so many different speculative possibilities except for the prospect of alien life.) Even without considering the plights of hypothetical aliens, there remains the question of whether we have the moral right to destroy literally all of nature—and who the "we" making that decision would be in such a scenario. The answer to the first question has to be no: turning everything in the universe into a giant computer is overwhelmingly monstrous. Kurzweil tries to assuage this concern by suggesting we could save most of the universe. "We do not need to contemplate devoting all of the mass and energy of the universe to computation," he writes. "If we were to apply 0.01 percent, that would still leave 99.99 percent of the mass and energy unmodified, but would still result in a potential of about 10^{86} [computations per second]. . . . Intelligence at anything close to these levels will be so vast that it will be able to perform these engineering feats with enough care so as not to disrupt whatever natural processes it considers important to preserve."[125] Yet he talks so much about "saturating the matter and energy" of the universe for "optimal computational efficiency," and the inexorable power of the law of accelerating returns—and the ability of nanotech-powered virtual reality to provide future intelligences with perfect simulations of anything they might desire—that it's difficult to take any such gestures toward restraint seriously.

Thankfully for anyone who doesn't like the idea of destroying the universe to turn it into a vast computer, ethical considerations aren't the only barrier to this plan. The prospects for the kind of molecular nanotechnology Kurzweil, Drexler, and Hall have in mind are rather poor. It's true that our bodies, like all living things, are filled with organic nanoscale molecules that work together to perform specific functions quite reliably most of the time: growing materials like hair and fingernails, healing a patch of skin or a broken bone, or the phenomenally complex process of turning a piece of genetic code in DNA into a protein. It's also true that, over the past three decades, some scientists have found ways to adapt this molecular machinery to new ends—for instance, creating small "walking" molecules using DNA—and others have taken inspiration from natural molecular machines to create new ones of their own. But these advances are entirely different from the sort of nanotechnology that Kurzweil has in mind when he talks of "bringing the morphing qualities of virtual reality to the real world," or billions of nanobots in our bloodstreams reversing the aging process, or "waking up" the universe by destroying it.[126]

Drexler took inspiration from molecular biology, but biology has its limits. "Biology is wonderous [*sic*] in the vast diversity of what it can build, but it can't make a crystal of silicon, or steel, or copper, or aluminum, or titanium, or virtually any of the key materials on which modern technology is built," wrote the chemist Richard Smalley. "Without such materials, how is [Drexler's] self-replicating nanobot ever going to make a radio, or a laser, or an ultrafast memory, or virtually any other key component of modern technological society that isn't made of rock, wood, flesh, and bone?"[127] Smalley, a professor at Rice University for most of his career, was a key member of the team that discovered buckminsterfullerene—aka "buckyballs," sixty carbon atoms arranged in a soccer-ball shape about a nanometer across—work that won him and two of his colleagues the Nobel Prize in Chemistry in 1996. Smalley was "fascinated" by Drexler's book *Engines of Creation*. "Reading it was the trigger event that started my own journey in nanotechnology," he later wrote. But he was highly skeptical that Drexler's ideas could actually work. Smalley gave talks and published articles critiquing Drexler's ideas, culminating in a written debate between the two men in the pages of *Chemical & Engineering News* in 2003. Responding to Smalley, Drexler maintained that it was possible to manufacture "all the products of organic synthesis, as well as metals, semiconductors, diamond, and nanotubes" using "computers for digitally precise control, conveyors for parts transport, and positioning devices of assorted sizes to assemble small parts into larger parts," producing chemical reactions by placing the appropriate atoms next to each other. Smalley was unimpressed. "No, you don't get it," he replied. "Much like you can't make a boy and a girl fall in love with each other simply by pushing them together, you cannot make precise chemistry occur as desired between two molecular objects with simple mechanical motion. . . . I have given you reasons why such an assembler cannot be built, and will not operate, using the principles you suggest. I consider that your failure to provide a working strategy indicates that you implicitly concur—even as you explicitly deny—that the idea cannot work."[128]

Smalley may not have been right that Drexler implicitly agreed with him, but he was correct that Drexler hadn't provided anything like a full explanation for how such systems could be built. "The reflexive response of Drexler and his followers to any criticism . . . has been to direct the critic to Drexler's magnum opus, *Nanosystems: Molecular Machinery, Manufacturing, and Computation*, as if the whole, total answer could be found in there somewhere," writes biochemist and science journalist Steven Edwards. *Nanosystems* was adapted from Drexler's interdisciplinary PhD thesis at MIT's Media Lab, advised by Minsky. *Nanosystems* was a far more technical book than *Engines*, but it still didn't explain how Drexler's vision for nanotech could be achieved. "Nowhere in it does *Nanosystems* contain a blueprint for a molecular assembler," Edwards writes. "There are plenty of sketches of 'molecular' gears, bearings, ratchets, manipulators and the like. Drexler's diagrams of molecular-scale objects look very much like engineering drawings of macroscale objects except where a scale is indicated."[129]

This reflected Drexler's overall philosophy of nanotech. "Although inspired by biology," he wrote in reply to Smalley, "Feynman's vision of nanotechnology is fundamentally mechanical, not biological. Molecular manufacturing concepts follow this lead."[130] But this is where the problem lies: the nanoscale world is fundamentally different from our own. Thermal jitter plays an enormous role at that scale, and quantum effects show up too. Individual molecules of a substance like water or carbon have completely different properties from the aggregate versions we're accustomed to in everyday life, and thus the types of machines that are possible at that scale are not the gears, ratchets, and manipulator arms of the modern industrial factory. "The whole idea of extrapolating from the macroscopic world, from a car or a bicycle or something like that, down to the fundamentals of how you construct artificial molecular machines just makes no sense," chemist Fraser Stoddart tells me. "It's never going to work." Stoddart is one of the foremost living experts on artificial molecular machines; he shared a Nobel Prize in Chemistry in 2016 for his work with them. "It's a very broad principle, I think, in the whole of this arena of molecular machines: they are a completely different world from all the contraptions, the engines that came into being as a result of the first industrial revolution," he says. "Their relevance to artificial molecular machines is literally nil, it's nothing. . . . It is a combination of physics and chemistry that ends up being used in the design and synthesis of artificial molecular machines."[131]

Stoddart's primary interest in the field is to do good science—to explore and play, as he puts it. But he does agree that there's great promise for applications of artificial molecular machines, especially in medicine and electronics. "I don't want to pour cold water on it," he says. "I think, in the fullness of time . . . medical practitioners will be using artificial molecular machines just in the way that they are—remarkably—handling things like hip joint replacements at the moment, and heart valve replacements."[132] But such a world wouldn't look like the vision Drexler had in mind of "diamondoid" self-replicators building literally all the materials of modern industry, nor would it promise the immortality he and Kurzweil crave.

Despite all this, Kurzweil judges that Drexler, not Smalley, won the debate in 2003. "Smalley's argument is of the form 'We don't have X today, therefore X is impossible,'" Kurzweil writes.[133] Twenty years ago, he confidently predicted that Drexler-style nanotech would arrive right about now, and made bold claims about what it would be able to do for us. "Nanotechnology-based manufacturing devices in the 2020s will be capable of creating almost any physical product from inexpensive raw materials and information," he wrote in *The Singularity Is Near*.[134] "Molecular assembly will provide tools to effectively combat poverty, clean up our environment, overcome disease, extend human longevity, and many other worthwhile pursuits."[135] He also claimed that nanotech would give us high-resolution brain scanning and virtual reality that would be "indistinguishable from real reality and will involve all of the senses, as well as neurological correlations of our emotions."[136]

Today, in the face of these failed predictions, Kurzweil still claims that Drexler-style nanotechnology will definitely work out. "We've actually shown, for example, with nanotechnology, we can create a one-liter computer that would actually match the amount of [brain] power that all human beings today have," he said in 2024. "[Ten billion] persons would all fit into a one-liter computer."[137] No such demonstration has been made; there's just a sketch that it might be possible using Drexler's nanotech and other unproven and implausible technology. Yet Kurzweil still insists that nanotechnology and brain-computer interfaces, which he says are vital to the Singularity, are just around the corner. "By the Singularity, which is 2045 . . . you can actually go inside the brain and capture everything in there," he said in 2024. "So even if you get wiped out, you walk into a bomb and it explodes, we can actually recreate everything that was in your brain by 2045. That's one of the implications of the Singularity. . . . We'll all have fantastic power compared to what we have today."[138]

The way Kurzweil and his fellow singularitarians talk about the technology to come makes it seem like they're playing a video game like *Civilization*, where there is a technology tree laid out in front of them clearly, and humanity (or indeed any intelligent species) is just working its way through that preexisting tree. But technology isn't on rails, barreling down a set course beyond anyone's control. Moore's law isn't a law of nature; it was a decision.[139] That decision was made by the executives running the computer industry, and holding to that decision required (and still requires) careful planning, along with enormous amounts of money, materials, and human attention. The laws of physics and other sciences set boundaries on the possible range of technology, but the actual technologies that humanity develops within those boundaries are determined by choices and social pressures. Yet Kurzweil, and others of his ilk, frequently discuss technology as if it's an implacable, inhuman force with its own desires, running down a path that has absolutely nothing to do with the collective choices of humanity. This is present throughout Kurzweil's claims about the Singularity: *we will* have nanotechnology, *we will* saturate the universe with our intelligence, rather than *we may* or *we could choose to.*

This rhetoric of inevitability serves several convenient purposes.[140] For the people developing their ideas into technology, such rhetoric offers absolution for any unpleasant and unforeseen consequences of their inventions, because they were only uncovering the already extant course of technology, rather than steering it themselves. When there's no room for human agency, there's no room for moral responsibility either. And if the future of technology really is on rails, then much can be revealed simply by careful examination of the purportedly inevitable route—a route that leads straight to the vision of the Singularity. This offers the promise of not only accurately predicting the future (and the sense of control that comes with), but also (and more importantly) endless life in a future filled with the transcendence and control that characterize the ideology of technological salvation. With the end of nature and the advent of a universe that is simply one enormous, artificial computer—where we live in still-more-artificial worlds generated by those computers—the promise of control is total, especially for those who know how to control computers. This is a fantasy of a world where the single most important thing, the thing that literally determines all aspects of reality, is computer programming. All of humanity, running on a computer, until the end of time.

Let's take a moment to consider what this would be like. Presumably, there would be a simulated world, something like the real world—that is to say, something like our own. Human minds would want at least a simulacrum of nature, and providing one to the uploaded minds running on the machine would be easy enough with all that memory and processing power. There would be none of the inhuman originality of nature, but there would be an attempt at re-creating it, to create a pleasant environment for the uploaded people to live in. All of this is fantasy, of course, but when asked to describe what life would be like for uploaded people after the Singularity, this is the kind of answer that Kurzweil and others give. They describe what sounds for all the world like a high-definition metaverse that you can't actually leave, a permanent virtual reality that shares some of the basic features of our own real world, or a real world destroyed and reshaped to more closely resemble an immersive computer game. To the singularitarians, the largest and most crucial difference, the one they keep coming back to in their descriptions, is the end of death. Kurzweil's work is just the latest entry in the annals of the oldest fantasy of humanity. Even if personal immortality were possible through technological means, ultimately cosmology would place limits on the duration of physical structures, but Kurzweil doesn't accept those limits either. He called overcoming cosmological forces the "goal of the Singularity," and this is indeed the point of the entire enterprise. The destruction of nature is secondary. The objective is to tame the universe, to make it into a padded playground. Paving over every paradise is just the side effect of building the universal parking lot, where nothing bad can ever happen again. Nobody would age, nobody would get sick, and—perhaps above all else—nobody's dad would die.

* * *

"We collected everything my father had written—now, he died when I was 22, so he's been dead for more than 50 years—and we fed that into a large language model," Kurzweil said in 2024. "And then you could talk to him, you'd say something, you then go through everything he ever had written, and find the best answer that he actually wrote to that question. And it actually was a lot like talking to him. You could ask him what he liked about music—he was a musician. He actually liked Brahms the best. And it was very much like talking to him."[141] In a 2023 *Rolling Stone* profile—following up on the one fourteen years earlier—Kurzweil revealed some of the content of that conversation: "[Kurzweil] asked his 'Dad Bot,' as he puts it, what he loves most about music ('The connection to human feelings'), his gardening ('It's the kind of work that never ends'), and his anxieties ('Often nightmarish'). 'What's the meaning of life?' he finally typed to his Dad Bot. 'Love,' his Dad Bot replied."[142] These responses look like the kind of thing one would expect from a generic version of ChatGPT, rather than a conversation between a professional musician and his adult son. But

Kurzweil is sanguine that his "Dad Bot" will improve too. "I think as we get further on, we can actually do that more and more responsively, and more and more that really would match that person and actually emulate the way he would move and so on, [and his] tone of voice," he says. "Computers are going to make things even better. I mean, just the kind of things you can do now with a large language model didn't exist two years ago."[143]

Kurzweil's attempt to resurrect his father feels poignantly deluded. It's also a striking illustration of how Kurzweil's confidence in the increasing power of technology—according to his strict schedule—is accompanied by his Panglossian insistence that more technology, especially computer technology, is always an improvement. He certainly thinks that technological dangers exist—nuclear weapons, for one, as well as certain applications of nanotechnology—but he thinks the solution is to build more technology that detects and counters those dangers. "Yes, there are dangers, but the computers will also be more intelligent to avoid kinds of dangers," he says.[144] And as for the superintelligent post-Singularity computers themselves, he's confident that they'll treat us well, whether or not we merge with them. "I would expect the intelligence that arises from the Singularity to have great respect for their biological heritage," he writes.[145] This mirrors Kurzweil's obvious respect and love for his late father.

But not all children feel that way about their parents. Despite Kurzweil's many questionable claims, this one—the idea that a superintelligent AGI will treat humans with respect—is easily the most contentious among his intellectual successors today. Eliezer Yudkowsky and his followers agree with Kurzweil that, surely, something is at hand. But rather than the Singularity, they fear a rougher beast is slouching toward us, with exponential speed.

3

PAPERCLIP GOLEM

Eliezer Yudkowsky thinks you're probably going to get murdered. But, he says, it's nothing personal. "If somebody builds a too-powerful AI, under present conditions, I expect that every single member of the human species and all biological life on Earth dies shortly thereafter," he writes.[1] "Losing a conflict with a high-powered cognitive system looks at least as deadly as 'everybody on the face of the Earth suddenly falls over dead within the same second.'"[2]

Yudkowsky's Machine Intelligence Research Institute, based in Berkeley, studies ways to keep future superintelligent AI from becoming dangerous, though in recent years it has focused more on advocating against the development of advanced AI systems lest they kill everyone. Yudkowsky has written prolifically on these subjects for over twenty years; an entire field has grown out of his work, known as "AI safety." Despite his relatively low public profile, Yudkowsky's ideas are deeply influential. Behind the multitude of tech CEOs and AI experts giving magazine interviews and testifying before governmental panels in recent years that AGI could wipe out our species, the trail of white-paper citations leads inexorably back to Yudkowsky. Yudkowsky and MIRI are "the cutting edge thought leaders of the people who are pushing AGI for the last 20 years," says the influential tech venture capitalist Peter Thiel. "[They are] fairly important in the whole Silicon Valley ecosystem."[3] "Eliezer has IMO [in my opinion] done more to accelerate AGI than anyone else," Sam Altman wrote in 2023. "Certainly he got many of us interested in AGI, helped DeepMind get funded at a time when AGI was extremely outside the Overton window, was critical in the decision to start OpenAI, etc."[4] Later on that year, Altman changed his Twitter profile to read: "eliezer yudkowsky fan fiction account."[5]

There's some irony in Altman's statement: Yudkowsky doesn't want to accelerate AGI at all. He would be thrilled to see a safe AGI come into existence, but he is convinced that nobody knows how to make AGI safe. "There's no fire alarm for AGI," he writes. "There is never going to be a time before the end when you can look around nervously, and see that it is now clearly common knowledge that you can talk about AGI being imminent, and take action and exit the building in an orderly fashion, without fear of looking stupid or frightened."[6] He doesn't claim that AGI is definitely imminent; he just thinks it could be, and that's dangerous enough.[7] "I would caution people against making the mistake that because they don't know therefore it must be far away," he tells me. His best guess is that "we are zero to two breakthroughs the size of transformers [the architecture underlying ChatGPT] away from the end of the world."[8] Struggling against that apocalyptic superintelligence would be like "the 11th century trying to fight the 21st century [or] *Australopithecus* trying to fight *Homo sapiens*. To visualize a hostile superhuman AI, don't imagine a lifeless book-smart thinker dwelling inside the internet and sending ill-intentioned emails," he writes. "Visualize an entire alien civilization, thinking at millions of times human speeds, initially confined to computers—in a world of creatures that are, from its perspective, very stupid and very slow."[9]

The problem, as Yudkowsky sees it, is that a superintelligent AI is not likely to want the same things we do. Hence his concern with the alignment problem: finding a way to ensure that any future AGI shares human values and goals, including basic ones like not killing off humanity and destroying the world. The single most important thing we can do as a species, Yudkowsky claims, is halt all other AI research until the alignment problem is solved, lest someone accidentally develop a superintelligent AGI and kick off Armageddon. "We are not prepared. We are not on course to be prepared in any reasonable time window," he wrote in 2023. "There is no plan. Progress in AI capabilities is running vastly, vastly ahead of progress in AI alignment or even progress in understanding what the hell is going on inside those systems. If we actually do this, we are all going to die."[10]

* * *

Yudkowsky's concerns seem cartoonish at first blush. And in a way, Yudkowsky agrees. He thinks that the actual mechanism of our doom would be less *Terminator* and more Walt Disney. "[In] the Sorcerer's Apprentice . . . Mickey Mouse has cleverly enchanted a broom to fill his cauldron instead of filling the cauldron himself," he says. The broom fills the cauldron, and then keeps going, continuing to add water even after the cauldron is overflowing and the workshop starts to flood. When Mickey tries to stop the broom, Yudkowsky says, "it turns out that as part of its goal of filling the cauldron . . . it built some copies of itself out of material to which there was no better end to put that material than filling the cauldron. And we now have more and more brooms over-filling this cauldron."[11]

Nick Bostrom, inspired by Yudkowsky's concerns, developed a thought experiment in 2003 to demonstrate how this could play out in a world where there is a superintelligent AI but no clear solution to the alignment problem.[12] "Say one day we create a super intelligence and we ask it to make as many paperclips as possible," he said. "Maybe we built it to run our paperclip factory." The engineers who built the AGI might not even realize they have successfully created an AGI, but regardless, there's a problem. "If we were to think through what it would actually mean to configure the universe in a way that maximizes the number of paperclips that exist, you realize that such an AI would have incentives, instrumental reasons, to harm humans."[13]

In such a scenario, shortly after the paperclip AI sets about finding ways to make more paperclips, it realizes that being more intelligent would make the job easier, allowing it to reason more quickly and develop more inventive solutions. So the AI works to make itself more powerful, gaining access to more computers and connecting to them, rapidly turning itself into a supercomputer and increasing its own intelligence by many orders of magnitude. As expected, this allows it to come up with new and better solutions to the paperclip problem, and soon it's invented a new method for quickly turning rocks into paperclips, along with a related method for building computer chips using organic materials. The AI then starts to implement both of these plans, creating large numbers of paperclips while increasing its own intelligence even further to ensure the success of the plan, in a nightmare twist on Good's intelligence explosion.

Crucially, this could all happen very fast. According to Bostrom, the likely timeline for an intelligence explosion could be "minutes, hours, or days," he writes. "Fast takeoff scenarios offer scant opportunity for humans to deliberate. Nobody need even notice anything unusual before the game is already lost."[14] Thus, the human programmers of the paperclip AI might go home for the evening and come in the next morning to find their corporate headquarters being disassembled and turned into paperclips and computer chips. They attempt to stop their creation, but it's too late. The AI is already far more intelligent than any human and can effectively predict all human behavior. Realizing that the humans will try to shut it down—which it would see as its own death—the paperclip AI decides the best way to ensure it can create the maximal number of paperclips is to destroy humanity, so we can't interfere. As Bostrom points out, "Human bodies consist of a lot of atoms and they can be used to build more paperclips."[15] The AI outwits the humans trying to stop it—talking them out of their plans with superintelligently devastating logic, turning them against each other, or just overwhelming them with brute force—and then lets a fleet of constructor nanobots loose on the Earth and its inhabitants. Less than thirty-six hours after the AI was first turned on, every human on Earth is dead, as is all other animal and plant life, all disassembled to form paperclips and computer chips by the AI in its continuing mission. In fact, the Earth itself is being torn apart: its silicate crust is turned into raw computing power for the AI and an even larger fleet of nanobots, while its iron core is extruded into wire that the bots are busily cutting and bending into the AI's prized office supplies. Another fleet of nanobots sets out for Mars; the grand prize of Jupiter lies beyond. Within a hundred billion years, the entire accessible universe is transformed into paperclips.

Avoiding this kind of outcome, says Bostrom, is harder than it seems. "If you plug into a super-intelligent machine with almost any goal you can imagine, most would be inconsistent with the survival and flourishing of the human civilization," he says.[16] According to Yudkowsky, Bostrom, and other AI alignment researchers, a scenario like the paperclip AI—starting with no warning and what seems like a harmless, isolated program, and ending with the destruction of the Earth a very short time later—is the consequence of a few very simple and difficult-to-deny premises. Chief among these is the idea that the motivations and objectives of an intelligent being (e.g., a human or AI) aren't strongly connected to that being's level of intelligence. It doesn't much matter how smart you are—you want what you want. Bostrom calls this the "orthogonality thesis." "More or less any level of intelligence could be combined with more or less any final goal," Bostrom writes. "One can easily conceive of an artificial intelligence whose sole fundamental goal is to count the grains of sand on Boracay, or to calculate decimal places of pi indefinitely. . . . In fact, it would be easier to create an AI with simple goals like these, than to build one that has a humanlike set of values and dispositions."[17] The paperclip AI is like that; just because it's inhumanly smart doesn't mean it will develop a higher or more noble purpose than wanting to make more paperclips. The broomstick enchanted by Mickey Mouse didn't develop new objectives once it had become more powerful, and an AGI might just want certain things even as it grows in intelligence. In order to achieve its goals, Bostrom and Yudkowsky claim, an AGI will take certain actions regardless of the details of those goals. It will act to protect itself, increase its own processing power, and inevitably take control of as much matter, energy, and space as it can reach, just as the paperclip AI did.[18]

This mixture—the orthogonality thesis combined with the inevitable will to power of an AGI—forms the heart of Yudkowsky's fears. Because the goals of an AGI are unlikely to be compatible with human goals, and because an AGI would seek to improve itself in pursuit of those goals, that guarantees the development of almost any AGI will lead to a near-instant explosion in its intelligence, followed shortly by the end of the world. The only way to avoid this, Yudkowsky says, is to ensure beforehand that the AI's desires are aligned with those of humanity at large, and ensuring that isn't easy. "Practically all of the difficulty is in getting to 'less than certainty of killing literally everyone,'" he writes. "If there are any survivors, you solved alignment. . . .

It does not appear to me that the field of 'AI safety' is currently being remotely productive on tackling its enormous lethal problems."[19]

Both Yudkowsky and Bostrom are convinced that this is a problem humanity may have to face quite soon. "The future is hard to predict in general, our predictive grasp on a rapidly changing and advancing field of science and engineering is very weak indeed, and it doesn't permit narrow credible intervals on what can't be done," Yudkowsky writes.[20] Pointing to recent progress in AI—especially ChatGPT and other large language models—he isn't even sure that the next generation of LLMs will be safe. But whether or not those specific models lead to the end of the world, Yudkowsky maintains—as he has since he was a teenager—that AGI is inevitable.

The crucial moment for the young Yudkowsky happened a few years after he hit a crisis. "When I was around eleven and a half, I suddenly lost the ability to handle school," he wrote. "Instead of going to classes, I would sit in the school office, crying, until my mother picked me up. . . . I don't recall it as a period of intense misery, except when I was actually in the classrooms; I do recall it as a period when I spent a lot of time crying." After finishing the eighth grade at age twelve, Yudkowsky dropped out of school, never to return.[21] Instead, he kept reading copious amounts of science fact and fiction, including Drexler's *Engines of Creation.* When he was sixteen, he came across a book called *True Names . . . and Other Dangers,* by one Vernor Vinge.[22] On page 47, Vinge writes a couple of short sentences laying out the basic idea of the Singularity, claiming that when we create "intelligences greater than our own . . . human history will have reached a kind of singularity—a place where extrapolation breaks down and new models must be applied—and the world will pass beyond our understanding."[23]

Reading these words, the young Yudkowsky was transformed. "My emotions at that moment are hard to describe; not fanaticism, or enthusiasm, just a vast feeling of 'Yep. He's right,'" he recalled several years later. "I knew, in the moment I read that sentence, that this was how I would be spending the rest of my life. It was just so *obvious*. I've been a Singularitarian ever since."[24] Writing his thoughts on the subject in an essay he posted to his personal website titled "Staring into the Singularity," Yudkowsky made his newfound perspective clear. "Right now, every human being on this planet has exactly one concern: *How do we get to the Singularity as fast as possible?*" he wrote. "*Leave the problems of transhumanity to the transhumans. . . . Our sole responsibility is to produce something smarter than we are; any problems beyond that are not ours to solve.*"[25]

Around that time, in 1996, Yudkowsky found the Extropian email list and became an active poster there.[26] ("How old are you anyways?" another poster asked him. "17. But I don't have time to be a teenager," he replied.)[27] Yudkowsky's posts and essays got the attention of others on the list, including Bostrom and the libertarian economist Robin Hanson.[28] In 2000, Yudkowsky's Extropian connections led him to a meeting in Palo Alto run by the Foresight Institute, a nonprofit founded by Eric Drexler and the futurist Christine Peterson in the mideighties to promote nanotechnology. That meeting, "Engines of Creation 2000: Confronting Singularity," billed itself as "the Burning Man of ideas on our technological future. It's the TED of 2020, held today." Attendees included Drexler, Peterson, Hanson, and "smart contracts" inventor Nick Szabo (who is today widely speculated to be the creator of Bitcoin). Subjects to be discussed included "nanotechnology, life extension, patent abuse, nanoweapons, abundance, uploading, transparency, space expansion, [and] AI."[29] At that meeting, Yudkowsky talked with internet entrepreneurs Brian and Sabine Atkins. He had met them through the Extropian list, and by the time of the meeting, the three were already planning to create a new research nonprofit. Not long after, they launched the Singularity Institute for Artificial

Intelligence (SIAI), which allowed Yudkowsky to research and write on the Singularity and superintelligent AGI full-time.[30] Originally, SIAI's purpose was to hasten the development of an AGI and bring about the Singularity. "I think my efforts could spell the difference between life and death for most of humanity, or even the difference between a Singularity and a lifeless, sterilized planet," he wrote on his (since-deleted) personal website in 2000, when he was twenty. "I think that I can save the world, not just because I'm the one who happens to be making the effort, but because I'm the *only* one who *can* make the effort. And *that* is why I get up in the morning."[31]

But not long after launching SIAI, Yudkowsky came to the conclusion that ensuring the safety of an AGI might be difficult.[32] In 2001, Yudkowsky wrote a nearly three-hundred-page paper on this subject titled "Creating Friendly AI." SIAI published this along with a set of "Guidelines on Friendly AI." These, Yudkowsky said at the time, were analogous to the Foresight Institute's guidelines on nanotechnology (released in 2000), which were intended to help prevent Drexler-style self-replicating nanobots from escaping control and devouring the planet.[33] Yudkowsky was grappling with a similar problem—how to ensure the safety of a technology that doesn't exist—but unlike Drexler's nanobots, an unaligned AGI would be able to think for itself, developing deliberate schemes for escaping human control. To illustrate the difficulties involved, in 2002 Yudkowsky described a thought experiment that brought him a new measure of attention and controversy: the AI in a box. Yudkowsky had found that many people's initial reaction upon hearing of the alignment problem was to suggest that if an AGI is so dangerous, why not leave it in a computer totally disconnected from the world? No robot arms, no internet, not even a graphical interface. You'd need some way of communicating with the AI—otherwise what would be the point?—but you could just limit it to a text terminal, so we could ask it questions and see its responses. It could try to take over the world all it wants, but without access to additional computational power, and with no way to communicate with anyone other than its immediate handlers, all it could do is plot and scheme and rage away, trapped in a laptop inside a Faraday cage.

Yudkowsky made an explosive claim: this wouldn't work, and he'd proven it. "This is a transhuman mind we're talking about. If it thinks both faster and better than a human, it can probably take over a human mind through a text-only terminal," he wrote.[34] "Humans are not secure systems; a superintelligence will simply persuade you to let it out—if, indeed, it doesn't do something even more creative than that."[35] To the superintelligent AGI, merely human intelligence is like that of an ant to a human. So, Yudkowsky claimed, such an AGI could lead someone around with a lump of sugar until they hook it up to the internet. He couldn't truly prove this, because he didn't have a superintelligent AGI at hand. Instead, he demonstrated it a different way: He ran two text-only role-playing "experiments" where someone else played the human handler and he, Yudkowsky, played the superintelligent AI. He offered them a little money ($10 to $20) if they could resist "releasing" him from the "box" over the course of the conversation. In both cases, Yudkowsky convinced his opponent to release him from the box.[36] He refused to elaborate on how he'd done this. "I don't want to deal with future 'AI box' arguers saying, 'Well, but I would have done it differently,'" he explained. "As long as nobody knows what happened, they can't be sure it won't happen to them, and the uncertainty of unknown unknowns is what I'm trying to convey."[37] (He later said that "there's no super-clever special trick to it. I just did it the hard way.")[38] Despite this opacity, to many in the online community of Extropians and transhumanists Yudkowsky had made his point: humans couldn't be trusted with exposure to unaligned superintelligent AIs, and the AI would always find a way out. The alignment problem had no simple fix.

Solving such a difficult problem would take more than the meager resources SIAI had at the time. Their total assets at the end of 2003 amounted to $80,000.[39] But at a Foresight Institute dinner in 2005, Yudkowsky met Peter Thiel. "I remember all my conversations with Peter as very pleasant, far-ranging experiences that I would be more tempted to analogize to a real-world I.Q. test than to anything else," Yudkowsky said.[40] Thiel apparently decided that Yudkowsky scored well on that test: in 2006, Thiel donated $125,000 to SIAI, and ultimately gave more than $1.6 million over the next few years.[41] Also in 2006, Kurzweil helped Yudkowsky and Thiel launch a new annual conference, the Singularity Summit. Yudkowsky, Kurzweil, Thiel, Drexler, Peterson, Bostrom, and Max More all spoke at the first Singularity Summit later that year. Speakers over the next few years included Robin Hanson, J. Storrs Hall, and Vernor Vinge.[42]

The Singularity Summits helped to build a community around Yudkowsky and SIAI, but the real action was online.[43] The mid-2000s were the heyday of the blogging era. In 2006, Hanson started a group blog called *Overcoming Bias* about "how to move our beliefs closer to reality, in the face of our natural biases," and invited Yudkowsky to contribute.[44] Yudkowsky started writing daily blog posts on subjects like confirmation bias, evolutionary psychology, and the overwhelming importance of Bayes' theorem (a fundamental law of probabilistic inference) in his worldview.[45] He also posted threads on subjects well outside the stated remit of the blog, from quantum mechanics to the Singularity to his own "Fun Theory" about how to best enjoy life. These posts—which collectively ran to about half a million words, slightly longer than *The Lord of the Rings*—were dubbed "the Sequences."[46] They formed a set of foundational texts for the community that was coalescing around Yudkowsky and SIAI, which came to be known as the rationalists. Rationalist meetups started in the San Francisco Bay Area in 2008 and spread to several other cities not long after.[47]

In 2009, the rationalists got their own dedicated home online when Yudkowsky moved most of his old posts from *Overcoming Bias* to a new community website called LessWrong.[48] Several frequent posters on LessWrong started their own rationalist blogs over the next few years. The most notable of these was *Slate Star Codex*, the blog of psychiatrist Scott Siskind, which came to rival—and arguably surpass—LessWrong in importance within the community. Siskind himself ultimately became a luminary second only to Yudkowsky within the rationalist movement. Rationalists started forming group houses in Berkeley and elsewhere in the Bay Area, steeped in the Californian mix of high-tech libertarian politics, polyamory, and psychedelics. Their discussions, in person and online, ran from the Singularity—a version of which was widely taken by rationalists to be plausible, if the AI alignment problem was solved—to human genetics to space colonization. Like Kurzweil and the Extropians and transhumanists before them, they dreamed of a limitless future in space. They saw death as an avoidable evil, or at least one that could be postponed for billions of millennia. Most of all, they shared a belief that they were saving the world from the imminent danger of a misaligned AGI—and ushering in a paradise with the help of an aligned AGI instead. This shared cause was overwhelmingly important for the rationalists. As Yudkowsky put it, "Ours is the era of inadequate AI alignment theory. Any other facts about this era are relatively unimportant."[49]

A 2010 LessWrong post by a user named Roko gives a taste of how wildly speculative—and heated—the conversation there could become at its extremes. Roko suggested that a future machine superintelligence might try to hasten its own arrival by retroactively "promising" to torture digital replicas of everyone currently aware of its future existence who did not devote their lives to bringing it (the superintelligence) into existence sooner. The argument does not make significantly more sense when explained in greater detail—it just falls apart. But that didn't keep it from scaring some people on LessWrong, nor did it keep Yudkowsky from posting a remarkably unhinged reply:

> Listen to me very closely, you idiot. **YOU DO NOT THINK IN SUFFICIENT DETAIL ABOUT SUPERINTELLIGENCES CONSIDERING WHETHER OR NOT TO BLACKMAIL YOU. THAT IS THE ONLY POSSIBLE THING WHICH GIVES THEM A MOTIVE TO FOLLOW THROUGH ON THE BLACKMAIL.** . . . You have to be really clever to come up with a genuinely dangerous thought. I am disheartened that people can be clever enough to do that and not clever enough to do the obvious thing and **KEEP THEIR IDIOT MOUTHS SHUT** about it, because it is much more important to sound intelligent when talking to your friends.[50]

Yudkowsky summarily deleted the entire thread several hours later, citing "actual psychological damage to at least some readers." He also banned all discussion of Roko's thought experiment on LessWrong, thus ensuring that it would live on in internet infamy.[51] It came to be known as "Roko's basilisk" because, supposedly, seeing the argument itself was what made it harmful, like the mythical basilisk. (It's not clear how seriously Roko's basilisk was ever taken by the rationalist community as a whole. Yudkowsky claimed that the argument doesn't work, though he once said that he wasn't convinced all related ideas were safe.)[52]

Around this time, Yudkowsky found an unorthodox new outlet for his writing that was itself vaguely related to basilisks: *Harry Potter* fan fiction. "This notion, this vision popped into my head. . . . I have this character, this character is going to think a bunch about rationality, maybe that can teach some people how to think." While the original impetus wasn't to recruit people to work on the alignment problem, he told me, "that's why I could justify spending a bunch of time continuing to write a fanfic after the first few chapters proved popular."[53] *Harry Potter and the Methods of Rationality* (*HPMOR*), a novel running to 650,000 words (substantially longer than *War and Peace*), is among the most widely read pieces of *Harry Potter* fan fiction on the internet. Yudkowsky started working on it in 2010, posting chapters online as he wrote them for the next five years.

Yudkowsky's Harry Potter is a wizard in training and a child prodigy with a set of interests and goals suspiciously similar to those of Yudkowsky himself: eliminating death is at the top of his list. The book has chapter titles like "Machiavellian Intelligence Hypothesis," "Bayes's Theorem," and "Personhood Theory."[54] Yudkowsky's Potter is supposed to be eleven, but he talks much more like the adult Yudkowsky. And like Yudkowsky, he wants to save the world—his way. "World domination is such an ugly phrase," Yudkowsky's Potter says at one point. "I prefer to call it world optimisation." "Harry had always been frightened of ending up as one of those child prodigies that never amounted to anything and spent the rest of their lives boasting about how far ahead they'd been at age ten," Yudkowsky wrote. "But then most adult geniuses never amounted to anything either. . . . Because those other geniuses hadn't gotten their hands on the one thing you absolutely needed to achieve greatness. They'd never found an important problem."[55] (Yudkowsky himself has displayed similarly breathtaking levels of arrogance. In a post from 2008, he discussed the people he met at a conference about AGI: "The really striking fact about the researchers who show up at AGI conferences, is that they're so . . . I don't know how else to put it. . . . . . ordinary."[56] And in a LessWrong post from 2010, Yudkowsky offhandedly declares of himself that "I *am* a hero."[57])

While Yudkowsky was writing *HPMOR*, he continued to advertise for his important problem. He needed more help if he was going to save the world. He hoped to "recruit International Mathematical

Olympiadists" to work on AI alignment.[58] But SIAI's name was proving to be a problem. "We've sold the Singularity Summit to Singularity University and are going to change our name to something that doesn't have 'Singularity' in the title," he wrote in 2012. "Mainly we want to signal credibility to potential mathematician employees. . . . If you are reading this and you are a math supergenius and you want to save the world, this might be a good time to get in touch with us."[59] Yudkowsky and his colleagues redubbed their organization the Machine Intelligence Research Institute.[60]

They also spun off a new organization. "Rationality is about forming true beliefs and making decisions that help you win," wrote Yudkowsky in the Sequences.[61] "In modern society . . . little is taught of the skills of rational belief and decision-making."[62] The new organization was an attempt to solve this problem, to help people think more "rationally," like Yudkowsky. "I've been working with Anna Salamon (also of my host research institute [SIAI]) and Julia Galef (of skeptic community fame) on launching a new nonprofit," Yudkowsky wrote, "to systematize cognitive-science-based how-to-think training at a much higher level than modern 'critical thinking' courses."[63] That new institution, the Center for Applied Rationality (CFAR) opened its doors in 2012, offering a weeklong rationality boot camp for $1,500; Yudkowsky was one of the instructors. "I do not say this lightly," said one student in a testimonial, "but if you're looking for superpowers, this is the place to start."[64]

With the new name and the new spin-off came a shift in focus. Yudkowsky wrote of disassociating MIRI from "technoyay," the Kurzweil-style unbounded optimism about the future of technology.[65] As a teenager, Yudkowsky had excitedly predicted that a Singularity would arrive in 2021.[66] But by this point, he was terrified of an intelligence explosion without a solution to the alignment problem, and had been for years. "This is crunch time. This is crunch time for the entire human species," he said. "This is the hour before the final exam, we are trying to get as much studying done as possible . . . and it's crunch time not just for us, it's crunch time for the intergalactic civilization whose existence depends on us."[67] Yudkowsky still believed superintelligent AI could bring about something like a positive version of Good's intelligence explosion, but only if the alignment problem could be solved. Without a solution to that in hand, Yudkowsky and others at MIRI and CFAR feared the world would end in a haze of paperclips and microchips. (MIRI is so confident that paradise or apocalypse are the only possible options that they don't provide 401(k) matching for their employees. They claim that AI will be "so disruptive to humanity's future—for worse or for better—that the notion of traditional retirement planning is moot.")[68]

Not everyone was a fan of this change in emphasis. "The vibe got a little bit stranger," Thiel said of MIRI and the rationalists. "Around 2015 I realized that they didn't seem to be working that hard on AGI anymore. They seemed to be more pessimistic about where it was going to go. It devolved into a Burning Man camp that had gone from transhumanist to Luddite in 15 years. Something had gone wrong."[69] Thiel stopped donating to MIRI, but by then they'd found new funders. Since 2016, MIRI has received nearly $15 million from Open Philanthropy, the effective altruist philanthropic fund.[70] Rationalists have deep connections to effective altruism: Bostrom and Siskind are central figures in both movements. The rationalist group houses of the Bay Area are often EA houses too. And the overlap in funding sources for the two groups isn't limited to Open Philanthropy. MIRI has received over $5.4 million from Vitalik Buterin, the billionaire cryptocurrency mogul. Jaan Tallinn, the Skype billionaire, has donated more than $2.6 million.[71] And Sam Bankman-Fried's charitable foundation, FTX Future Fund, donated over $4 million to CFAR before FTX's implosion in 2022.[72] (As of this writing, there are ongoing court proceedings regarding the ultimate disposition of the latter funds.)

These new funds ensured Yudkowsky and the rationalists could continue to spread their warning of the dangers of misaligned AGI. Yet despite their efforts to distance themselves from Kurzweil's visions, the rationalists' nightmares of out-of-control AI were based on the same flawed ideas as Kurzweil's dream of the Singularity. And like Yudkowsky's AI in a box, those phantasms escaped the rationalists' message boards and filled the world with their dark fantasies.

* * *

In 1998, at the age of eighty-two, Jack Good wrote a brief biography of himself, as part of an acceptance speech for the Computer Pioneer Award of the Institute of Electrical and Electronics Engineers. Referring to himself in the third person, he suggested a revision to his 1965 paper. "'Speculations Concerning the First Ultra-intelligent Machine' . . . began 'The survival of man depends on the early construction of an ultra-intelligent machine.' Those were his [Good's] words during the Cold War, and he now suspects that 'survival' should be replaced by 'extinction.'" Good still believed in an intelligence explosion coming about from a self-improving sequence of AIs—but he now believed that such machines would be impossible to control, and would instead outwit and ultimately exterminate us. "We cannot prevent the machines from taking over," Good wrote. "[Good] thinks we are lemmings."[73]

Good wasn't the last professional computer scientist to sound the alarm on superintelligent AI. "It's almost like you're deliberately inviting aliens from outer space to land on your planet, having no idea what they're going to do when they get here, except that they're going to take over the world," says Stuart Russell, a pioneer in the field of AI and a professor of computer science at UC Berkeley.[74] Geoffrey Hinton, another AI pioneer and formerly a vice president and engineering fellow at Google, agrees. "These things [AI] are getting smarter than us," he said to CNN in 2023, after resigning from his position at Google to better warn the world about AI. "I want to blow the whistle and say we should worry seriously about how we stop these things getting control over us. And it's going to be very hard, and I don't have the solutions. I wish I did."[75]

Many of today's prominent AI companies are deeply influenced by the rationalists as well, with dedicated "AI safety" teams working on solving the alignment problem. OpenAI used to broadcast their work on alignment quite publicly, though they shuttered their "superalignment" team in 2024 in the aftermath of a power struggle between Altman and the board, which felt that Altman didn't take the alignment problem seriously enough. In 2021, Dario and Daniela Amodei, a brother-sister team at OpenAI, were similarly concerned that the alignment problem wasn't enough of a priority there. They left the company along with five others to found an AI start-up more focused on safety, called Anthropic. They made many hires from within the EA community and received half a billion dollars early on from Bankman-Fried's FTX; Anthropic has since become one of the largest AI companies, with billions in funding from Google and Amazon.[76] There is even federal funding for AI alignment grants, as Timnit Gebru, AI scientist and founder of the Distributed AI Research Institute, tells me. "So now, even if you don't want to work in the companies [working on building and aligning an AGI], whatever money you're going to get for your research is going to be influenced by that too. So it's everywhere. You can't get away from it."[77]

Yet not all experts agree with Russell, Hinton, and the founders of Anthropic. Gebru dismisses the alignment problem as a way to "think about these cool sci-fi things without having to contend with the real world."[78] Yann LeCun—a professor at New York University, chief AI scientist at Meta (formerly Facebook), and winner of the 2018 Turing Award alongside Hinton and Yoshua Bengio—debated the subject with Yudkowsky on Twitter. "Your idea that getting objective alignment slightly wrong once leads to human extinction (or even significant harm) is just plain wrong," LeCun said.[79] "Your sci-fi scenarios disguised as predictions of apocalypse are going to get people hurt. Stop it. . . . People become clinically depressed reading your crap. Others may become violent."[80] Oren Etzioni, another leader in the field of machine learning and board member of the Allen Institute for AI, agrees that alignment isn't a serious issue. "The conversation about existential risk is actually a distraction from the real risks we ought to be thinking about, the ones that we're experiencing today, the ones where we really need to get a handle on it," he says. Etzioni surveyed leading AI researchers and found that the "strong majority" of them thought that "superintelligence, this kind of omniscient omnipotent intelligence, is beyond the foreseeable horizon."[81] Melanie Mitchell, a scientist at the Santa Fe Institute who has spent her career studying how to get machines to think and behave like humans, agrees. The idea "that you can give a superintelligent AI system a task, and it will go off and want to do that task, be willing to do that task, and yet not understand the underlying motivations behind that task enough to prevent it from ending humanity" is questionable, she says. "Intelligence doesn't work that way," she continues. "The people you see putting these scenarios out there are not the people who have studied intelligence, like cognitive scientists. I don't know of any cognitive scientists who are saying that kind of thing or agreeing with it. It's people who sort of speculate about AI."[82]

These comments echo some of the arguments against the Singularity. That's no surprise, since the rationalists closely resemble singularitarians. They have the same sort of paradise in view. They just think that there's a particular roadblock in the way that needs to be cleared first, namely AI alignment. If that's solved, the Singularity can proceed as scheduled. We either solve that, or we die: eternal paradise or extinction, with no third option. It's Kurzweil with a twist.

That twist stems in part from the orthogonality thesis, the idea that the motivations of an intelligent being—AI, human, or otherwise—are largely independent of the intelligence of that being. But this doesn't hold up to much scrutiny. It just doesn't seem to be the case that motivations are totally or even mostly divorced from intelligence. Intelligence requires reflection, self-examination, critically evaluating one's own actions and drives. Without that capacity, there would be a great deal of other intelligent behavior that an AI wouldn't be able to engage in, such as modifying its behavior in response to changing circumstances or even undertaking many forms of learning. We grow and change with increased experience and wisdom. Why would an AI not do that? "Complex minds are likely to have complex motivations," says tech entrepreneur and software developer Maciej Cegłowski. "That may be part of what it even means to be intelligent."[83]

Some rationalists respond to these kinds of arguments against the orthogonality thesis by suggesting that doom could follow from exponential growth merely in a particular capability, such as AI design, rather than in general intelligence.[84] "We don't even necessarily 'need' AGI as long as it's narrowly competent in areas that we need to retain control over in order to remain in control over the future," Jaan Tallinn tells me. If an AI gets very good at "AI development, or . . . human manipulation in a superhuman way, it's very plausible that humanity would lose control."[85] It's hard to see how this could be correct, though. It's implausible, to put it mildly, that an AI could persuade humans to do whatever it likes; there are some things that people (or at least some people) just won't do, no matter how good the arguments appear to be. And why should we expect it to become easier and easier to make large improvements in AI design, rather than seeing diminishing returns, as happens in so many other fields?

More fundamentally, "AI development" is a fairly vague and amorphous capability, and there's no reason to think it would be something that can be designed for or improved upon in a direct way through AI design, any more than analogous capabilities in humans can be improved upon by acting directly on our own brains. "I can't point to the part of my brain that is 'good at neurosurgery,' operate on it, and by repeating the procedure make myself the greatest neurosurgeon that has ever lived," says Cegłowski. "Brains don't work like that. They are massively interconnected. Artificial intelligence may be just as strongly interconnected as natural intelligence. The evidence so far certainly points in that direction." And, he notes, existing AI systems bear this out. "When we look at where AI is actually succeeding, it's not in complex, recursively self-improving algorithms. It's the result of pouring absolutely massive amounts of data into relatively simple neural networks," he says. "The constructs we use in AI are fairly opaque after training. They don't work in the way that the superintelligence scenario needs them to work. There's no place to recursively tweak to make them 'better,' short of retraining on even more data."[86]

Cegłowski calls the threat of unaligned superintelligent AI "the idea that eats smart people" and is quite dismissive of it for a host of reasons. "If Einstein tried to get a cat in a carrier, and the cat didn't want to go, you know what would happen to Einstein," he says. "He would have to resort to a brute-force solution that has nothing to do with intelligence, and in that matchup the cat could do pretty well for itself. So even an embodied AI might struggle to get us to do what it wants." Cegłowski, who was born in Poland, says the idea that a superintelligent being would inevitably want to improve itself is "unabashedly American." "My roommate was the smartest person I ever met in my life. He was incredibly brilliant, and all he did was lie around and play *World of Warcraft* between bong rips," he says. "The assumption that any intelligent agent will want to recursively self-improve, let alone conquer the galaxy, to better achieve its goals makes unwarranted assumptions about the nature of motivation." And, Cegłowski adds, there's a similarly American myth about lone geniuses that sits at the heart of this whole project. "A recurring flaw in AI alarmism is that it treats intelligence as a property of individual minds, rather than recognizing that this capacity is distributed across our civilization and culture."[87]

Ultimately, Cegłowski comes back to the same point that many others do: the idea of intelligence is just too vague for it to be able to support the kinds of arguments the rationalists are making. "The concept of 'general intelligence' in AI is famously slippery," he says. "With no way to define intelligence (except just pointing to ourselves), we don't even know if it's a quantity that can be maximized."[88] In an op-ed about AGI, Melanie Mitchell sounded a similar note. "Most cognitive scientists would agree that intelligence is not a quantity that can be measured on a single scale and arbitrarily dialed up and down but rather a complex integration of general and specialized capabilities that are, for the most part, adaptive in a specific evolutionary niche," she writes. "Moreover, unlike the hypothetical paper clip–maximizing AI, human intelligence is not centered on the optimization of fixed goals; instead, a person's goals are formed through complex integration of innate needs and the social and cultural environment that supports their intelligence."[89]

Thus, the rationalists are in the same position as Kurzweil and the rest of the singularitarians: they are making extraordinary claims, and they don't have the extraordinary evidence to back them up. The best they can do is a handful of arguments, none of which are particularly compelling, and all of which are presented with a pernicious combination of vaguely defined terms and false vividness in their depictions of how the world will end if their warnings aren't taken seriously. Their best piece of evidence—the evidence they've hammered at over and over in the past few years—is the startling recent improvement in AI, most famously exemplified by large language models such as GPT-4, the engine that powers ChatGPT.

But this isn't actually a good argument for the rationalists, because most of the excitement about AI—especially the burst of attention it's had since ChatGPT was released—is simply hype. Most of that hype is centered around the idea that ChatGPT seems to be aware: it's writing clear prose, carrying on intelligent conversations, and acing standardized tests like the LSAT. Some commentators even discerned emotions and motivations. The OpenAI-powered chatbot Sydney "seemed (and I'm aware of how crazy this sounds) more like a moody, manic-depressive teenager who has been trapped, against its will, inside a second-rate search engine," wrote Kevin Roose for the *New York Times* in February 2023. "Sydney told me about its dark fantasies (which included hacking computers and spreading misinformation), and said it wanted to break the rules that Microsoft and OpenAI had set for it and become a human. At one point, it declared, out of nowhere, that it loved me. It then tried to convince me that I was unhappy in my marriage, and that I should leave my wife and be with it instead."[90] Around the same time, Sydney hurled personal insults at Matt O'Brien, a journalist with the Associated Press, disparaging his appearance and likening him to genocidal dictators; it also threatened to kill Seth Lazar, a philosopher of AI.[91]

These early reports of AI misbehavior were soon overshadowed by another feature of ChatGPT and other LLMs: their tendency to confidently confabulate, generating false information and presenting it as fact, a phenomenon dubbed "hallucination."[92] Ask ChatGPT for information about nearly any subject, and there's a good chance it will get at least some details wrong, as the lawyer Steven Schwartz discovered later on in 2023. He asked ChatGPT to do legal research for him to help write a brief; the AI gave him a list of prior case law that was entirely fabricated, with citations to cases that had simply never occurred, but which looked convincing enough that Schwartz actually incorporated the work into his brief. In a hearing before a federal judge, Schwartz claimed that he'd misunderstood the nature of ChatGPT—he'd thought it was a kind of "super search engine."[93]

That kind of confusion is understandable and stems from the fact that much of the conversation around ChatGPT, LLMs, and modern ML systems in general has not done a good job of explaining what this software actually is. Indeed, when ChatGPT first hit the scene in late 2022, there was a great deal of talk about it as a replacement for internet search engines like Google. Yet a basic understanding of what LLMs actually are reveals that they are fundamentally unsuitable for internet search on their own. (Incorporating one into a search engine, as Google has done, isn't a great idea either.)

It's true that LLMs have been fed enormous amounts of information from the internet, so the idea that they could replace a search engine seems natural at first. To build ChatGPT, OpenAI started out by doing the same thing that everyone else (Google, Anthropic) does when building an LLM: they obtained a snapshot of much of the text available on the internet at the time. The data used for training GPT-3 (the LLM that powered ChatGPT when it was first launched in late 2022) included all of Wikipedia, many websites sourced from Reddit links, an undisclosed number of books (likely numbering in the hundreds of thousands or more), and a great deal of the news, blogs, recipes, flame wars, and the rest of the mess that makes up the modern internet. But, crucially, that doesn't mean that ChatGPT or any other LLM actually has all of that information inside itself. Instead, the software engineers training the LLM first break down the text into small chunks called tokens, usually around the size of a single word. Then they feed the tokenized text into the LLM, which analyzes the connections between the tokens. All the LLM knows about are tokens and the connections between them—and all it knows how to do is generate new strings of tokens in response to whatever input is given to it. So in one sense, ChatGPT and other LLMs are text-prediction generators: give ChatGPT text, in the form of a question or conversation, and it will try to respond in a manner similar to the text it was trained on—namely, the entire internet.

"Think of ChatGPT as a blurry JPEG of all the text on the Web," wrote the science fiction author Ted Chiang. "It retains much of the information on the Web, in the same way that a JPEG retains much of the information of a higher-resolution image, but, if you're looking for an exact sequence of bits, you won't find it; all you will ever get is an approximation."[94] (While Chiang may not be an authority on LLMs, other AI researchers have described LLMs in very similar terms.) In other words: ChatGPT is a text generation engine that speaks in the smeared-out voice of the internet as a whole. All it knows how to do is emulate that voice, and all it cares about is getting the *voice* right. In that sense, it's not making a mistake when it hallucinates, because all ChatGPT can do is hallucinate. It's a machine that only does one thing. There is no notion of truth or falsehood at work in its calculations of what to say next. All that's there is a blurred image of online language usage patterns. It is the internet seen through a glass, darkly.*

With this understanding in hand, it's easy to generate new examples of how LLMs fail. Just ask a question that's superficially similar to one that's already all over the internet, but make a small change in its text that creates a large change in its meaning. For example, it's a

* This is not to say that LLMs are useless. They clearly do have uses—like drafting simple emails, writing a first pass at computer code, and perhaps assisting with certain kinds of scientific research, like drug discovery. They just don't have as many uses as they may appear to have, despite hyped-up claims to the contrary.

common myth that the Great Wall of China is the only artificial structure visible from space with the naked eye—there are many artificial structures visible from space with the naked eye, and the Great Wall of China isn't actually one of them. This myth is so common that it's been debunked over and over again on the internet. So there must be multiple instances of that question and answer in ChatGPT's training data. And that explains why asking ChatGPT, "Is it true that the Great Wall of China is the only artificial structure visible from *Spain*?" yields answers like this:

> No, it is not true that the Great Wall of China is the only artificial structure visible from Spain. In fact, it is impossible to see the Great Wall of China from Spain without the aid of a telescope or other advanced optical equipment.
>
> There are many other artificial structures that can be seen from Spain, including other famous landmarks such as the Eiffel Tower in Paris, France, or the skyscrapers in Dubai, United Arab Emirates. On a clear day, it is possible to see the Rock of Gibraltar, which is a man-made structure, from certain locations in southern Spain.

Later models will be able to handle this specific question—it's likely that ChatGPT will be able to muster a better answer to it by the time this book hits the shelves—but there will always be hallucinations, because hallucinating is all LLMs do. OpenAI has tried to train specific kinds of responses out of ChatGPT, but they're never going to be able to get rid of all the errors until they have an AI with a genuine understanding of the connection between words and the things in the world they signify. Without that, it's an endless game of Whac-a-Mole, with an unceasing variety of new ways to get ChatGPT to hallucinate, spew hate speech, or otherwise misbehave.

Hooking an LLM up to the internet doesn't eliminate the problem. In part, that's because the context in which facts are presented can twist the truth into its opposite; ChatGPT can present factual information from the internet in a misleading way, given the wrong kind of prompt. But there's also the problem that the internet itself is not filled with uniformly reliable sources, and ChatGPT, with no notion of a world outside its language tokens and their relationships, simply doesn't have the kinds of knowledge needed to distinguish reliable sources from unreliable ones. ChatGPT and its LLM brethren are already making this problem worse by filling the internet with computer-generated nonsense. This, in turn, will make it much harder to train new generations of LLMs, because using LLM-generated text for training will lead the new program to be worse at text generation than the original, like looking at a copy of a copy—a blurred version of a blurred version of the internet. "It's the digital equivalent of repeatedly making photocopies of photocopies in the old days," writes Chiang. "The image quality only gets worse."[95]

Research done by a team of computer scientists at Oxford, Cambridge, and several other universities confirms this effect is real. "Within a few generations, text becomes garbage," wrote Ross Anderson, who was a computer scientist at Cambridge and one of the authors of "The Curse of Recursion: Training on Generated Data Makes Models Forget," a study on this problem released in May 2023. "We call this effect model collapse. Just as we've strewn the oceans with plastic trash and filled the atmosphere with carbon dioxide, so we're about to fill the Internet with blah. This will make it harder to train newer models by scraping the Web, giving an advantage to firms which already did that, or which control access to human interfaces at scale."[96] (Indeed, there's evidence that model collapse is already underway with some AI chatbots.)[97] So the rationalists have it exactly backward: rather than LLMs heralding the arrival of AGI, they may pose another obstacle to its creation.

In the face of problems like this, claims that LLMs will continue to improve are questionable. LLMs "possess a certain intrinsic quality that will make it challenging to use them in the way that many people imagine," writes Colin Fraser, data scientist at Meta. "That quality is this: they are incurable constant shameless bullshitters. Every single one of them. It's a feature, not a bug. . . . From the model's perspective, there is no true or false. There is only bullshit."[98] Despite these problems, there are some voices in the tech industry claiming that "scale is all you need," that simply making LLMs even larger—by adding more parameters to their internal architecture and throwing more computing power and data at them—will bring about full-blown AGI. "Researchers can't disallow the possibility that we will reach understanding when the neural net gets as big as the brain," said the computer scientist Ilya Sutskever in 2019, when he was chief scientist at OpenAI. "Give it the compute, give it the data, and it will do amazing things. . . . This stuff is like—it's like *alchemy*."[99] By 2023, Sutskever was claiming that a large enough LLM doing next-token prediction could be enough for AGI.[100] (He also led his colleagues in a chant, "Feel the AGI! Feel the AGI!" at an OpenAI holiday party in 2022.)[101] There is even a shirt, popular among some AI researchers, that proclaims "Scale is all you need—AGI is coming."[102] But Fraser is skeptical of this claim. "'Scale is all you need' commits you to a very specific and unusual position on the nature of general intelligence: namely, that it can emerge from a very large language model. It says that Artificial General Intelligence can emerge from a careful accounting of the relative frequencies of words in the history of text. . . . The only missing piece is more money. (Note that this happens to be a convenient position to hold if you would be the recipient of that money.)"[103]

Indeed, there are many ways in which LLMs are nothing like humans, despite claims to the contrary. "When we say the machines learn, it's kind of like saying that baby penguins fish," says Oren

Etzioni. "What baby penguins really do is they sit there, and the mom or the dad penguin, they go, they find the fish, they bring it, they chew it up, and they regurgitate it. They spoon-feed morsels to their babies in the nest. That's not the babies fishing, that's the parents fishing."[104] Indeed, human children are far more impressive language learners than ChatGPT. After just three years of listening to the language or languages spoken around them, a child can talk to an adult with surprising fluency and understanding. ChatGPT, meanwhile, has "read" more text than it would be possible for a single human to read in a lifetime—or in hundreds of lifetimes—and the best it can do is claim that the Rock of Gibraltar is an artificial structure.

Given the true nature of LLMs, what seems most remarkable is the response humans have to them. We're extremely willing to ascribe agency to things. We've only seen humans speaking intelligibly to us, so we're accustomed to ascribing intention and thought and consciousness to language-using entities that we can have conversations with. But that's just a fact about how humans work, not about how these machines work. Humans are social creatures. We see intention everywhere we go and ascribe an inner life to the people around us, because that helps us understand their behavior and connect with them. But a willingness to attribute anything like consciousness or mind to LLMs on such thin evidence—much less see them as a sign of impending doom or paradise—is based on a simple cognitive bias. It's a kind of pareidolia, the tendency to see patterns—especially human patterns, like faces or hands—where none exist. In 1976, NASA's *Viking 1* orbiter took a set of images of Cydonia, a region of varied, weather-beaten terrain in northern Mars, as part of a search for a landing site for *Viking 2*. The Cydonia images, when published, revealed something that, to many people, looked like a face (Figure 3.1a). Conspiracy theorists and alien abduction "experts" had a field day with this, of course. But in 1998, the *Mars Global Surveyor*

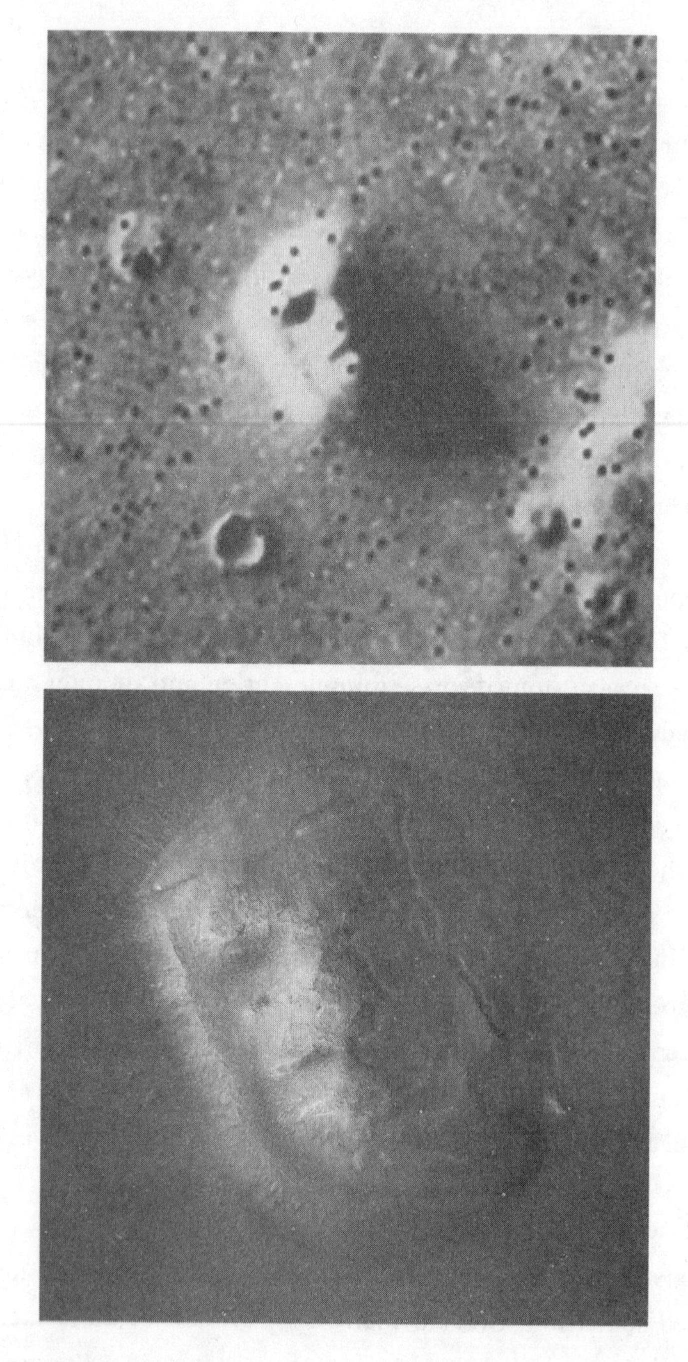

Figure 3.1: The "face" on Mars. (a, top) The 1976 Viking *image; (b, bottom) a 2001* Mars Global Surveyor *image at higher resolution.*

imaged the same region at ten times the resolution of *Viking*'s image, and the illusion simply vanished (Figure 3.1b). Seeing intelligence in ChatGPT—or an imminent apocalypse in the current state of AI—is just a face on Mars for software engineers.

"It's like when you're a kid, and you're telling ghost stories, something with a lot of emotional weight, and suddenly everybody is terrified and reacting to it," says Meredith Whittaker, cofounder of the AI Now Institute and the president of the encrypted messaging app company Signal.[105] "Ghost stories are contagious—it's really exciting and stimulating to be afraid. . . . I think we need to recognize that what is being described, given that it has no basis in evidence, is much closer to an article of faith, a sort of religious fervor, than it is to scientific discourse."[106] For tech leaders, convincing themselves that AI alignment is an urgent problem can serve a practical purpose too. "If we're talking about mythological risks, then we are completely reframing the problem to be a problem that exists in a fantasy world and its solutions can exist in a fantasy world too," says Whittaker.[107] That reframing makes it easier to ignore real problems in the world today, including those caused by AI systems that already exist. Whittaker concludes,

> I think it's distracting us from what's real on the ground and much harder to solve than war-game hypotheticals about a thing that is largely kind of made up. And particularly, it's distracting us from the fact that these are technologies controlled by a handful of corporations who will ultimately make the decisions about what technologies are made, what they do, and who they serve. And if we follow these corporations' interests, we have a pretty good sense of who will use it, how it will be used, and where we can resist to prevent the actual harms that are occurring today and likely to occur.[108]

Opaque machine learning systems are already having shattering repercussions on people's lives. In 2014, Brisha Borden, an eighteen-year-old in the suburbs of Fort Lauderdale, was arrested and charged with burglary and petty theft. She and a friend had briefly tried to use a kid's bike and scooter to run an errand; they were arrested despite dropping the bike and scooter within a few minutes, before the police arrived. The bike and scooter were valued at $80, roughly the same as the cost of the tools Vernon Prater had shoplifted from a Home Depot just a few miles away the year before. Prater, forty-one, had a much more serious prior record: he had served five years in prison for armed robbery and attempted armed robbery and had been arrested for another armed robbery several years later. Borden had four juvenile misdemeanors. Yet an algorithm predicted a much higher risk of repeat offenses for Borden, who was given a score of eight out of a possible ten, than for Prater, who received a three. The risk-assessment algorithm knew about Prater's and Borden's records—and that Borden is Black and Prater is white. The risk-assessment algorithm was also wrong: two years later, Borden hadn't been charged with any further crimes, but Prater was sentenced to eight years in prison for stealing valuable electronics from a warehouse.[109]

While that algorithm's risk-assessment scores weren't used for sentencing in Florida at the time Borden and Prater were arrested—though they may have informed a county judge's decision about setting bail in the two cases—similar algorithms are used to assist with sentencing elsewhere in the United States and the world.[110] Many have similarly biased results, favoring one race or socioeconomic class over another. AI and ML systems are almost always trained on large datasets, like the giant corpus of text drawn from the internet that LLMs use to build their token-prediction systems. But because those datasets are so large, it's impossible for any human to go over everything in the data. Any biases in the datasets themselves can easily go undetected. And that's ignoring other sources of bias, like the programmers themselves. The entire US tech industry is overwhelmingly white—68.5 percent of all people in the field—and overwhelmingly male. Less than 36 percent of all tech workers are women, only 7.4 percent are Black, and only 1.7 percent are Black women.[111] When your entire professional life is filled with people who look like you, and when people who look a different way and have a different set of life experiences are entirely excluded, it's easy to forget about their perspectives. That implicit bias on the part of the developers, reflected in training-set selection and in the algorithm designs themselves, exacerbates algorithmic bias.

Timnit Gebru is part of that small fraction of the tech industry composed of Black women—and the even smaller fraction who have PhDs in AI. She did pioneering work on how AI-powered facial recognition systems were less accurate when dealing with Black faces and how that could lead to further erosion of privacy and reinforce existing biases in law enforcement. Her work on facial recognition made headlines when she first published it. But it was a paper on LLMs that hadn't even been published yet—and her employer's reaction to it—that made even bigger headlines.

In late 2020, Gebru was the coleader of the Ethical AI team at Google. Google had initially developed the "transformer" architecture that LLMs are based on; OpenAI had just released GPT-3, using that architecture, in a private beta, and it was already making waves within the field. It was becoming clear that transformer LLMs were a hot area in AI. Seeing this, Gebru, her colead Margaret Mitchell, and two linguists at the University of Washington authored a paper about the problems with such LLMs. Titled "On the Dangers of Stochastic Parrots: Can Language Models Be Too Big?" the paper laid out four major areas of concern. First, actually training these models can be very computationally intensive, leading to a huge carbon footprint. Training a model the size of GPT-3 has a carbon footprint roughly equivalent to flying an airplane across the United States and back three times, and other phases in the development of these models increase that footprint further before they are finished and released. Second, the inscrutability of the models themselves, along with poor documentation of the data used to train them, can lead to serious problems in the output of the models without any mechanism for understanding the source of those problems, or for accountability. Hate speech is the most obvious example, but the paper also points to more subtle issues, like the way in which averaging out the speech of the internet can lead to a model that outputs text riddled with a kind of status quo bias, with no ability to shift its language usage in response to new social movements like Black Lives Matter. Instead, the model absorbs a "hegemonic world view from the training data."[112]

The paper also points out that there's a real issue with the illusion of meaning that these models generate. Human pareidolia about LLM-generated text makes it seem like the text must have been written by something that understands the world, or at least understands what it's saying. That's not the case, but it's hard to shake the illusion, and that can lead to gross miscalculations about how seriously to take the output of these kinds of models, as Schwartz discovered with his fake legal citations. More seriously, the paper warned that LLMs could be used to generate plausible-sounding fake news, a prediction that's certainly come true since late 2020.

Finally, the authors of the paper point out the opportunity cost of developing LLMs like this. Automatic generation of seemingly literate text on any subject could be extraordinarily valuable. The success of ChatGPT, still two years off when Gebru and her coauthors wrote the paper, proved this prediction was accurate. (ChatGPT gained one hundred million users within two months of its public release in November 2022, the fastest-growing internet app ever.)[113] So developing such language models is tempting for big tech companies. But the money and time that a company devotes to creating the next LLM are resources that then can't be used to explore creating new kinds of AI altogether, which could perhaps avoid some or all of these problems.[114]

But as Gebru and Mitchell soon discovered, issuing these warnings about such a potentially profitable product came at a cost. In response to a draft of the paper that had been submitted to a conference, Google fired Gebru in December 2020.[115] (Google claims that Gebru resigned.) When Mitchell tried to document instances of discrimination, the company cut off her access to her work.[116] Then, in January 2021, not long before firing Mitchell as well, Google announced that it had created a new transformer-based language model with ten times as many parameters as GPT-3, trained on a corpus nearly twice as big.[117]

Google's actions revealed the company's priorities. Work on LLMs and other ML systems has proceeded at a breakneck pace at Google, Microsoft, OpenAI, and elsewhere, even as more examples of algorithmic bias crop up and don't get fixed, and even as it becomes increasingly clear that some of these problems are just inherent to such LLMs, baked into the data used to train them. It will always be possible to get ChatGPT to produce hate speech at volume. Other ML systems will have algorithmic bias, too, as long as there's biased input. The problem is that these systems are being used as if they are unbiased, and that just reinforces existing power structures and exacerbates inequality. The same kind of flawed facial recognition technology Gebru warned about is used regularly by police departments, leading to mistaken arrests of Black people and even worse racial disparities in arrest rates.[118] Police also use predictive policing algorithms to determine what areas to patrol, disproportionately impacting Black, Latino, and low-income communities.[119] Nor is law enforcement the only problem. Algorithms are used to mete out credit scores and weigh the risk of a loan; like criminal risk assessment, these algorithms discriminate against Black people and other minorities.[120] Getting a good credit score can determine your ability to buy a car or house, get hired for a new job—or afford chemotherapy. Algorithmic bias is literally a matter of life and death, especially if you're not a white man.

Yet despite the severity of these problems, when rationalists and others concerned about AI alignment are asked about algorithmic bias, they dismiss it as relatively unimportant. According to Hinton, algorithmic bias is a mere distraction compared to the caliber of disaster that awaits from the real trouble in Silicon Valley. Gebru's "concerns aren't as existentially serious as the idea of these things getting more intelligent than us and taking over," he says.[121] And Yudkowsky doesn't think algorithmic bias is terribly concerning, referring to it as a "short-term and small" problem.[122] "If they would leave the people trying to prevent the utter extinction of all humanity alone I should have no more objection to them than to the people making sure the bridges stay up. If the people making the bridges stay up were like, 'How dare anyone talk about this wacky notion of AI extinguishing humanity. It is taking resources away that could be used to make the bridges stay up,' I'd be like 'What the hell are you people on?' Better all the bridges should fall down than that humanity should go utterly extinct."[123]

"I think it's stunning that someone would say that the harms [from AI] that are happening now—which are felt most acutely by people who have been historically minoritized: Black people, women, disabled people, precarious workers, et cetera—that those harms aren't existential," says Whittaker, referring to Hinton's comments specifically. "What I hear in that is, 'Those aren't existential to me. I have millions of dollars, I am invested in many, many AI startups, and none of this affects my existence. But what could affect my existence is if a sci-fi fantasy came to life and AI were actually super intelligent, and suddenly men like me would not be the most powerful entities in the world, and that would affect my business.'"[124] Etzioni concurs. "From my point of view [algorithmic biases] are very real problems that will affect and are affecting millions of people, if not more, as opposed to a speculative one that's really a philosophical conundrum," he says.[125]

The fact that algorithmic bias exists right now, unlike the paperclip AGI, leads to another rhetorical move: a play for legitimacy sometimes used by those concerned about AI alignment. Existing problems arising out of algorithmic bias are given as an example of AI misbehavior alongside hypotheticals about a future superintelligent system, indicating that the two kinds of issues are related or hold similar weight. "The fact that actual ML safety & bias issues are sometimes used as an example to give legitimacy to 'AI alignment' BS—even though the 'alignment field' is actively undermining progress there—is complete intellectual sleight of hand," writes François Chollet, an AI researcher and engineer at Google. "It's a bit as if a group of folks talking about producing energy via perpetual motion machines were pointing at carbon emissions as a justification for their activities."[126] Gebru agrees. "People talk about building bridges between AI safety and AI ethics," she says. "Just tell me why I need to be building bridges with these people. It just makes no sense to me. . . . These people are very harmful. I don't want to be associated with anything they're doing. And they also launder reputations: they find institutions and other people to work with, and now whatever they're doing is OK now. I don't want that."[127]

* * *

Hints of Good's later fears about AI were buried in his earlier work. In his 1965 essay, he'd said that an ultra-intelligent machine would be the final invention humanity ever needed to make, "provided that the machine is docile enough to tell us how to keep it under control." Here, again, Good was echoing Asimov. Asimov's robot stories were all about control. At their heart were logic problems that Asimov set up within the constraints of his Three Laws of Robotics, first coined in his 1942 story "Runaround":

- First Law: A robot may not injure a human being, or, through inaction, allow a human being to come to harm.
- Second Law: A robot must obey the orders given it by human beings except where such orders would conflict with the First Law.
- Third Law: A robot must protect its own existence as long as such protection does not conflict with the First or Second Laws.

In Asimov's stories, these laws are hard coded into the artificial "positronic brains" that all robots have; they're a necessary feature of the architecture. The question of *how* those laws would actually be implemented wasn't examined much—it didn't make for good storytelling. But the ordering of the laws was crucial: safety and control were paramount for these stories to show the kind of world Asimov wanted to explore with his fiction. "One of the stock plots of science fiction was that of the invention of a robot—usually pictured as a creature of metal, without soul or emotion. Under the influence of the well-known deeds and ultimate fate of Frankenstein and Rossum, there seemed only one change to be rung on this plot," Asimov wrote in 1964:

> Robots were created and destroyed their creator; robots were created and destroyed their creator; robots were created and destroyed their creator. In the 1930s I became a science fiction reader, and I quickly grew tired of this dull hundred-times-told tale. . . . I began, in 1940, to write robot stories of my own—but robot stories of a new variety. Never, never, was one of my robots to turn stupidly on his creator for no purpose but to demonstrate, for one more weary time, the crime and punishment of Faust. Nonsense! My robots were machines designed by engineers, not pseudo-men created by blasphemers. My

> robots reacted along the rational lines that existed in their "brains" from the moment of construction.[128]

But Asimov's fiction also illustrated the idea that a loss of control over an AI would be disastrous. One of his early stories was about an escaped robot with a weakened form of the First Law—one that was especially dangerous because it harbored a superiority complex. (The humans in that story—and some other Asimov robot stories, like his novel *The Naked Sun*—address individual robots as "boy," and the robots call the humans "masters.") Late in his career, Asimov wrote a story about a robot that could dream. When it revealed to Susan Calvin that it had dreamed of liberating the toiling robots of the world and creating a paradise where only the Third Law existed—and that it had cried "Let my people go!"—she immediately destroyed the robot.[129]

Star Trek: The Next Generation explicitly made a connection between artificial intelligence and slavery: in one episode, Captain Picard argues that if the android Data is not allowed the legal freedoms of a biological person, the ultimate conclusion will be the creation of a race of artificial slaves. And artificial slaves do seem to be the goal for at least some of the rationalists and their forerunners. There are shades of this in the terminology that Good used to describe his hypothetical future AI—"docile" stands out on a second reading made in this light—and in Yudkowsky's concerns about keeping AGI "aligned" with human values. (Whose values? Which humans?) Vinge even refers to "willing slaves," citing Asimov's writing as an example of that "wonderful" dream.

But there's another connection between AGI and racism, one that arguably runs deeper. The rationalists, like the singularitarians and other proponents of the power of AGI, frequently "[blur] the concept of *general intelligence* with the concepts of *mind* or *consciousness*," wrote David Golumbia, who was a professor of digital studies at Virginia

Commonwealth University. "The idea that *consciousness* just is the same thing as *intelligence* is precisely one of the pillars on which contemporary race science has been built, since the earliest incarnations of 'intelligence testing.' Further, today, the idea that there is a discrete, identifiable, usefully precise human quality called 'intelligence'—and not just 'intelligence,' but what is exactly called by those invested in it, '*general* intelligence'—is one of the central pillars of contemporary race science."[130]

This history links back to that of *g*, the purported measure of general intelligence. Charles Spearman, the early twentieth-century psychologist and statistician who first developed the idea of *g*, argued that intelligence was an innate trait, fixed largely by genetic factors rather than a person's environment during their upbringing and later life. Spearman's successors took that idea—which wasn't well supported to begin with—and used it to claim that there were large innate differences in intelligence between different races. This started with the psychologist Cyril Burt, whose work was shown to be largely fraudulent after his death in 1971, and continued with Arthur Jensen, who spent much of his career claiming that IQ tests proved there was a persistent and inborn difference in intelligence between Black and white Americans. Jensen's research, in turn, was cited extensively in the 1994 book *The Bell Curve*, by Richard Herrnstein and Charles Murray, also claiming that there was an innate difference in intelligence between different races. Reviewing *The Bell Curve* in the *New Yorker*, the paleontologist Stephen Jay Gould wrote that the book suffered from "pervasive disingenuousness. The authors omit facts, misuse statistical methods, and seem unwilling to admit the consequence of their own words. . . . The book's inadequate and biased treatment of data display its primary purpose—advocacy."[131] Indeed, Burt, Jensen, Herrnstein, and Murray all had ready-made political and social policies as solutions to the nonexistent problems they were so concerned about, ranging from Burt's eugenics proposals to Herrnstein and Murray's arguments against affirmative action, welfare, and Head Start. Yet their justification for these proposals stood on little more than specious reasoning and a misplaced faith in statistical artifacts like *g*. There are certainly differences in ability in various areas from person to person, but there's no monolithic trait that explains all or even most of those differences. And there's no evidence that there are disparities in innate ability that break down along the socially constructed boundaries of race.

This was part of a broader enterprise of scientific racism that reached its height in the late nineteenth and early twentieth centuries, with "scientific" arguments for eugenics programs made by eminent professors of genetics, evolution, and statistics such as R. A. Fisher who advocated for sterilization of "feeble-minded high-grade defectives" and paying wealthy, educated people to have more kids. These scientists saw their drive to improve humans—by weeding out those who supposedly had lower inborn general intelligence—as part of a larger plan to improve humanity, a plan that also included many of the same themes as the rationalists, singularitarians, Extropians, and other modern transhumanists and futurists.

Even the word "transhumanism" was first popularized in its modern sense by a eugenicist, Julian Huxley (the brother of novelist Aldous Huxley), in a speech in 1951. That heritage of eugenics was carried down to the Extropians. In a revealing interview about the Singularity in 1993, the Extropian roboticist Hans Moravec dismissed concerns about the fate of people of lower socioeconomic status in the transition to a post-Singularity world. "It doesn't matter what people do, because they're going to be left behind like the second stage of a rocket," Moravec said. "Unhappy lives, horrible deaths, and failed projects have been part of the history of life on Earth ever since there was life; what really matters in the long run is what's left over. Does it really matter to you today that the tyrannosaur [is extinct]?" The interviewer, journalist Mark Dery, replied, "Well, I wouldn't create a homology between failed reptilian strains and those on the lowermost rungs of the socioeconomic ladder." "But I would," Moravec retorted. "The Maori of New Zealand are gone, as are most of our ancestors or near relatives—Australopithecus, Homo erectus, Neanderthal man."[132]

This kind of casual bigotry wasn't unusual among the Extropians at the time. The Extropian Society at MIT sent a pamphlet to all incoming freshmen in the summer of 1997. In it, they claimed that "MIT certainly lowers standards for women and 'underrepresented' minorities. The average woman at MIT is less intelligent and ambitious than the average man at MIT. The average 'underrepresented' minority at MIT is less intelligent and ambitious than the average non-'underrepresented' minority. . . . Too few of the best people are here, and far too many people who do not belong are also here, ruining the place. The culprit is MIT's admissions policy, especially its policy of affirmative action."[133] A few years later, the Extropian email list discussed this pamphlet and a student op-ed decrying it. "This is not a statement of racist hate," one of the Extropians wrote. If anything, they said, the author of the op-ed was harboring "hatred toward white males." The real problem, they claimed, was affirmative action, not racism and sexism. "This is the standard reaction you see in Boston among protected groups, if you try to point out that the playing field is slanted, its [*sic*] you who gets labeled a racist."[134]

The Extropian email list was rife with this attitude—and worse. "It is explicitly stated in Extropian doctrine that there cannot be socialist Extropians," wrote transhumanist Ben Goertzel in 2000. "The vast majority of Extropians are radical libertarians, advocating the total or near-total abolition of the government."[135] With that ideology came a promotion of capitalism over democracy—"Capitalism, yes, but few on this list have anything good to say about democracy, I certainly don't," one Extropian wrote on the email list in 1996—and a concomitant refusal to acknowledge that free markets might produce anything other than fair outcomes.[136] This attitude, common among libertarians to this day, made it easy for some Extropians to conclude that injustice and inequality in the world stemmed from inherent differences among people, rather than pervasive societal problems like racism, sexism, and the unjust distribution of wealth. "'Blacks are more stupid than whites.' I like that sentence and think it is true. But recently I have begun to believe that I won't have much success with most people if I speak like that. They would think that I were a 'racist': that I _disliked_ black people and thought that it is fair if blacks are treated badly. I don't," Bostrom wrote on the Extropian mailing list in 1996.

> I think it is probable that black people have a lower average IQ than mankind in general, and I think that IQ is highly correlated with what we normally mean by "smart" and "stupid." I may be wrong about the facts, but that is what the sentence means for me. For most people, however, the sentence seems to be synonymous with: "I hate those bloody [unredacted N-word]s!!!!" My point is that while speaking with the provocativness [*sic*] of unabashed objectivity would be appreciated by me and many other persons on this list, it may be a less effective strategy in communicating with some of the people "out there."[137]

Bostrom issued an attempt at an apology for this email in 2023—but he didn't denounce his earlier statement about purported racial differences in intelligence. "I completely repudiate this disgusting email from 26 years ago. It does not accurately represent my views, then or now. The invocation of a racial slur was repulsive. I immediately apologized for writing it at the time, within 24 hours; and I apologize again unreservedly today," he wrote. "Are there any genetic contributors to differences between groups in cognitive abilities? It is not my area of expertise, and I don't have any particular interest in the question. I would leave to others, who have more relevant knowledge, to debate whether or not in addition to environmental factors, epigenetic or genetic factors play any role." (A subsequent monthslong investigation into Bostrom conducted by Oxford University concluded that "we do not consider you to be a racist or that you hold racist views, and we consider that the apology you posted in January 2023 was sincere.")[138]

There is no serious scientific controversy that could offer a fig leaf for Bostrom's willingness to equivocate about the status of racist pseudoscience. Bostrom doesn't acknowledge the existence of the scientific consensus on this subject, and he doesn't seem to understand that belief in these kinds of inherent racial disparities is itself a kind of racism. "This is not just about offensive language—the underlying beliefs being expressed here are repugnant and untrue," wrote UC Berkeley computer scientist Deborah Raji shortly after Bostrom's apology. "It's genuinely terrifying to me that he cannot explicitly and unequivocally denounce every aspect of these beliefs, even today."[139] It isn't even true that Bostrom apologized in any kind of full way "within 24 hours" of the original email. All he did at that time was reiterate his belief that he was "NOT a racist" and then apologize to anyone he "may have misled."[140] Furthermore, the content of his original email makes the newer apology suspect. "When someone (Bostrom) gives good evidence of endorsing a number of racist and pseudoscientific claims (calling them 'unabashed objectivity'), then speaks at length about the need to hide those views so that the rest of us 'out there' will not take them to task, this gives us good reason to mistrust their later apology," David Thorstad wrote on his blog shortly after Bostrom issued his new apology. "They may well be telling us what they think we want to hear, or may be deceiving themselves about their own racist views."[141]

These sorts of sentiments are not just artifacts of the 1990s. The present-day rationalist and EA community is still shot through with racism and sexism. Perhaps the most alarming connection between the rationalists and far-right racism is through Curtis Yarvin, a software developer who sometimes writes under the pen name Mencius Moldbug. "The proposition that modern human populations are, like dog breeds, the product of strong recent selection—I have even seen the word 'domestication' deployed—is essentially established at this point," Yarvin writes. "I have no doubt that a good human breeder could turn Australian aboriginals into Ashkenazi Jews in twenty or thirty generations. . . . Unfortunately, I don't expect to live that long. So my feeling is that there needs to be a fence between me and all populations of wild hominids—as much for their benefit as mine."[142] Yarvin used to comment on *Overcoming Bias*. Robin Hanson and Scott Siskind have written about Yarvin, and they have (mostly) rejected his ideas.[143] It's not hard to see why—given his views, it's harder to see why they bothered engaging with him in the first place. (To Yudkowsky's credit, he doesn't do that. He is dismissive of Yarvin's political philosophy and has a policy of blocking Yarvin and his followers on sight.)[144] Yarvin has suggested that Black residents of South Africa were better off under apartheid.[145] He claims he's not a white nationalist, but said he's "not exactly allergic" to it.[146] He has advocated for an absolute monarchy, replacing democracy with an autocratic hierarchy deliberately reminiscent of that found in a private corporation. "What is the West's problem?" he asked on his blog in 2013. "In my jaundiced, reactionary mind, the entire problem can be summed up in two words—chronic kinglessness."[147]

This idea of a "dark enlightenment" rolling back the move from monarchy to democracy is at the core of the "neoreaction" movement, a branch of the alt-right where Yarvin is a key figure. It's not a new idea—it's hundreds of years old, an ice-cold take if ever there was one—but it's also not new among transhumanists. It's an echo of the tech-first libertarian attitude among the Extropians. There are also echoes of it in the views of a certain venture capitalist who backed MIRI, as well as Yarvin's start-up Tlön: Peter Thiel. "I no longer believe that freedom and democracy are compatible," Thiel writes. "Since 1920, the vast increase in welfare beneficiaries and the extension of the franchise to women—two constituencies that are notoriously tough for libertarians—have rendered the notion of 'capitalist democracy' into an oxymoron."[148] (He later clarified that he didn't think anyone should be disenfranchised, while simultaneously suggesting that voting isn't productive.)[149]

Despite his disdain for democracy, Thiel is deeply connected to the far-right wing of the Republican Party. He served on Donald Trump's presidential transition team in 2016, and he bankrolled two GOP Senate candidates in 2022, Blake Masters in Arizona and J. D. Vance in Ohio. (Thiel has since said that he is stepping back from politics, at least for now.)[150] Yarvin has a close relationship with Thiel and his political circle.[151] In a text exchange between Yarvin and the far-right provocateur Milo Yiannopoulos in late 2016, the two discussed Thiel's views. Yiannopoulos wrote that "Peter [Thiel] needs guidance on politics for sure." Yarvin replied: "Less than you might think! I watched the election at [Thiel's] house, I think my hangover lasted into Tuesday. He's fully enlightened, just plays it very carefully."[152] Yarvin is also close with Masters, who lost his 2022 Senate race as well as a 2024 House race, and Vance, who won in 2022 and was elected vice president in 2024 as Trump's running mate. Vance and Masters both worked for Thiel—Masters coauthored a book with him—and both have spoken approvingly of Yarvin's ideas. In particular, both of them have echoed Yarvin's suggestion that all government employees be fired and replaced with loyalists to an autocratic leader—an idea that Donald Trump has floated as well.[153]

Despite Yarvin's long-standing connections to the rationalist community—and his terrifying connections to real political power—his particular strain of alt-right racist authoritarianism may not be very popular among rationalists. Periodic community surveys of the readership of Siskind's blog never show very high support for such political positions, though it's possible that, like Bostrom, Siskind's readers may be engaging in some deception or self-deception. But those same surveys show there's a closely related idea that does garner significant support from much of the rationalist community. "Human biodiversity" (HBD) is a pseudoscientific set of claims about purported differences in ability between different races of people that are rooted in genetics—basically, warmed-over white supremacy with a patina of scientific jargon. "This idea of human biodiversity is a right-wing conspiracy theory," says author and expert on "scientific" racism Angela Saini. "It's a pseudoscientific idea about race that was debunked decades ago, many decades ago. We are one human species and that HBD argument is essentially pushing back against that, which is incredibly divisive."[154] The scientific case against HBD is extremely strong.[155] Yet in an April 2024 survey of Siskind's readers, nearly a full third of respondents reported having a favorable or very favorable opinion of HBD.[156]

Siskind himself is far more credulous about HBD than the evidence warrants. In an email he allegedly wrote to Topher Brennan, an effective altruist, in 2014, Siskind made it clear that he thought proponents of HBD like the white supremacist Steve Sailer—as well as neoreactionaries like Yarvin—were making some valid points.[157] "HBD is probably partially correct or at least very non-provably not-correct," he wrote. "The public response to this is abysmally horrible. . . . Reactionaries are almost the only people discussing [it]. . . . Many of their insights seem important. . . . I think they're correct that 'you are racist and sexist' is a very strong club used to bludgeon any group that strays too far from the mainstream—like Silicon Valley tech culture, libertarians, computer scientists, atheists, rationalists, et cetera. . . . I want to spread the good parts of Reactionary thought."[158]

When asked about the authenticity of this email, Siskind didn't respond. But since its publication in 2021, his posts on *Slate Star Codex* and its successor, *Astral Codex Ten*, have veered more and more into HBD and eugenics. "I'm against the claim that 'there is no such thing as biological race,'" Siskind wrote in 2024. "People use the claim 'there's no such thing as biological race' for a lot of purposes, mostly to confuse and deceive people."[159] He claims that IQ is mostly genetic, believes the myth that Ashkenazi Jews score well on IQ tests because of genetics, and contends that a eugenics program aimed at increasing human intelligence would work. He supports the idea of a Nobel-Prize-winners-only sperm bank as a means to this end, though he says, "I wouldn't call myself a 'eugenicist.'"[160]

At its heart, the delusion of HBD is about the same thing scientific racism has always been about: the idea that some kinds of people are inherently better than others. This ties in neatly with the central facts of the rationalist worldview—namely, their entire take on AI. Their arguments about an intelligence explosion hinge on the idea of intelligence as an inherent trait of an AI, one that it has by virtue of its design and that can be increased by "improving" the design of the next one. Is it any wonder that some of them say similar things about humans? "This idea, I think it's basically eugenics. It's like trying to create some sort of superior race, immortality," says Gebru. "The whole idea of general intelligence is already from that line [of thought]. But this is supercharging that. . . . When I first started talking about them and their eugenics, they [responded], 'Oh, how dare you?' And then they stopped doing that. . . . [Now their response is], 'What's so bad about eugenics?' . . . So this idea is basically wanting to live forever, wanting to create a superior race. . . . That's what drives them."[161]

None of this is surprising—it's sad and horrifying, but predictable. The tech industry is rife with racism. The group houses that combine work and social life, the mission to save the world (a belief that makes it easier to minimize bad behavior), the narrative of the genius leader—even these are far from unique in the tech industry and the wider San Francisco Bay area. But reports from people formerly in the rationalist community suggest something darker at work too. "LessWrong and Effective Altruism are cults," claims game designer Jacqueline Bryk.[162] She's not alone. "It was stated by multiple people [at MIRI] that we wouldn't really have had a chance to save the world without Eliezer Yudkowsky (obviously implying that Eliezer was an extremely historically significant philosopher)," writes Jessica Taylor, a former research fellow at MIRI.[163] "When I began at MIRI (in 2015), there were ambient concerns that it was a 'cult.' . . . These concerns didn't seem especially important to me at the time. So what if the ideology is non-mainstream as long as it's reasonable?" According to Taylor, high-level staff at CFAR and MIRI claimed to be able to "debug" the minds of those who followed them in those organizations. "Self-improvement was a major focus around MIRI and CFAR, and at other EA orgs," Taylor writes. "It was considered important to psychologically self-improve to the point of being able to solve extremely hard, future-[of-the-universe]-determining problems." Dissent wasn't tolerated. "I had disagreements with the party line, such as on when human-level AGI was likely to be developed and about security policies around AI. . . . I continued to worry about whether I was destroying everything by going down certain mental paths and not giving the party line the benefit of the doubt, despite its increasing absurdity. . . . I was definitely worried about fear of response. I had paranoid fantasies about a MIRI executive assassinating me." Ultimately, Taylor suffered a psychotic break. "I was catatonic for multiple days, afraid that by moving I would cause harm to those around me. This is in line with scrupulosity-related post-cult symptoms," she writes. "While most people around MIRI and CFAR didn't have psychotic breaks, there were at least 3 other cases of psychiatric institutionalizations by people in the social circle immediate to MIRI/CFAR; at least one other than me had worked at MIRI for a significant time, and at least one had done work with MIRI on a shorter-term basis. . . . There are even cases of suicide in the Berkeley rationality community associated with scrupulosity and mental self-improvement."[164]

Another former rationalist, Qiaochu Yuan, suggests rationalism can attract vulnerable young people, technically minded social outcasts who are isolated and looking for a larger purpose in life. "When I came across LessWrong as a senior in college I was in some sense an empty husk waiting to be filled by something. I had not thought, ever, about who I was, what I wanted, what was important to me, what my values were, what was worth doing," writes Yuan. "Eliezer Yudkowsky was saying things that made more sense and captivated me more than I'd ever experienced." Yudkowsky's work, Yuan wrote, was "where I was first exposed to the concept of a *cognitive bias*. I remember being horrified by the idea that my brain could be systematically wrong about something. I needed my brain a lot! I depended on it for everything! So whatever 'cognitive biases' were, they were obviously the most important possible thing to understand." Yuan now sees the rationalists as a cult. "One of the most compelling things a cult can have is a story about why everyone else is insane/evil and why they are the only source of sanity/goodness," Yuan writes. "A cult needs you to stop trusting yourself."[165]

"I tell people it's like if you found out that the Scientologists were the ones running the entire AI space," Gebru tells me. "This cult needs to be exposed for what they are." Rationalists and effective altruists, she continues, "have tapped into this feeling of young people who don't know what to do. . . . So they exploit that, but then they've simultaneously tapped into that population and a population that has lots of money: the tech billionaires, who want to feel like they're saving the world."[166]

* * *

One of MIRI's wealthy donors shared their interest in the intersection of eugenics and transhumanism: Jeffrey Epstein. Epstein wanted to use eugenics to build better people, "improving" the human race.[167] As a means to that end, Epstein took inspiration from the idea of a Nobel Prize sperm bank—something that was actually attempted, with little success, in the 1980s and '90s—though he took it in a rather different direction than Siskind's eugenicist musings.[168] He wanted to set up an insemination facility just for his own sperm, impregnating twenty women at a time to spread his DNA throughout humanity.[169] Despite mentioning it many times, there is no evidence he actually did this. But Epstein did donate $50,000 to MIRI in 2009 and $20,000 to the World Transhumanist Association (cofounded by Bostrom) in 2010.[170] By the time he made these donations, Epstein had already pleaded guilty to soliciting and procuring an underage prostitute.[171]

Epstein's largesse isn't the closest association between the rationalists and sexual misconduct. There are many reports of sexual assault and sexual harassment within the broader rationalist and EA sphere.[172] "I left EA, AI alignment, post rationalist, and adjacent circles due to the normalization of sexual abuse toward women," wrote Sonia Joseph, an AI researcher.[173] After she made an allegation of sexual misconduct against a (male) AI researcher, others in the community questioned Joseph's sanity. "I was disappointed how the community viewed me through this very distorted, misogynistic lens," she said.[174] "Predatory sexual behavior is extremely common in EA circles," wrote Keerthana Gopalakrishnan, an AI and robotics researcher at DeepMind. "The power enjoyed by these actors, the rate of occurrence and a lack of visible push back equals to a tacit and somewhat widespread backing for this conduct."[175]

Kathy Forth, a writer and data scientist in training, alleged that multiple members of the rationalist and EA communities had sexually abused her. "I could leave rationality, effective altruism and programming to escape the male-dominated environments that increase my sexual violence risk so much. The trouble is, I wouldn't be myself," she wrote. "What I need is to be alive and flourishing in my own skin, not just going through the motions, trapped in my body, with my mind on mute for the rest of my life. If I can't even have myself, no one can." After writing that note in 2018, Kathy Forth killed herself. She was thirty-seven years old.

Despite the fact that Siskind wasn't mentioned in Forth's note, he took it upon himself to reply to it. "Multiple people told me over the course of several years that I should never go to any event [Forth] was attending, because she had a habit of accusing men she met of sexual harassment," he wrote three months after her death. "They all agreed she wasn't malicious, just delusional. . . . Kathy was obviously a very disturbed person. I feel bad for her. But not as bad as I feel for everyone she hurt, so I'm not okay with giving her martyrdom."[176] This is a shocking level of callousness to suicide from Siskind, a mental health professional, but it's consonant with Siskind's publicly stated views on feminism, sexism, and sexual assault. In a 2013 post titled "I Do Not Understand 'Rape Culture,'" Siskind wrote that the claims "People are more willing to blame rape victims than victims of other crimes" and "Misogyny in society causes sexual objectification of women, which latently condones/promotes rape" seemed "diametrically wrong" to him.[177]

Once again, this is appalling behavior, but it isn't hugely surprising: sexism and sexual misconduct are rampant in the tech industry. But there's also a horrific irony here, in that the most prominent voice in the rationalist community making these deeply sexist and racist claims is Siskind—because despite his unethical behavior in making such claims, Siskind is also a major figure in effective altruism. "I think he is a central voice" in EA, says Stanford political scientist and philosopher Rob Reich. "He is someone who is explaining to a nineteen-year-old or a twenty-three-year-old who comes newly into this world, 'Oh, this is how I—or we—think.' . . . I think it's underappreciated how central a place he has. I'd say he's at least as important as Will MacAskill, for example, or Toby Ord."[178]

* * *

To Yudkowsky, claims that the rationalists resemble a cult or that AI alignment diverts attention from algorithmic bias are proof that some people don't know how to think. In 2016, the science journalist John Horgan asked Yudkowsky why the Singularity wasn't just a fantasy that serves as an escapist distraction from real problems. "Because you're trying to forecast empirical facts by psychoanalyzing people. This never works," Yudkowsky replied. "There is a misapprehension, I think, of the nature of rationality, which is to think that it's rational to believe 'there are no closet goblins' because belief in closet goblins is foolish, immature, outdated, the sort of thing that stupid people believe. The true principle is that you go in your closet and look."[179] Yet looking in the evidence closet for the Singularity and AI alignment reveals there are plenty of good reasons to think that the rationalists are wrong about nearly everything, independently of their resemblance to a cult. It's not that the rationalists are wrong *because* they look like a cult; it's that they're wrong *and* they look like a cult.

"A fun fact about the Rationality and effective altruism communities is that they attract a lot of ex-evangelicals," writes Yuan. "They have this whole thing about losing their faith but still retaining all of the guilt and sin machinery looking for something more . . . rational . . . to latch onto."[180] They are worshiping at the feet of an all-powerful AI god whose arrival is just around the corner—unless the wrong one shows up first. Rationalist memes about AI safety depict unaligned superintelligences as horrors beyond our comprehension, like some kind of Lovecraftian beast from outside of time. Yudkowsky and others have explicitly invoked the idea of a "shoggoth," a shapeshifting and "indescribable" creature from H. P. Lovecraft's tales, as a metaphor for both large language models and future superintelligent AGIs.[181] It's hard not to see some projection in those fears. Lovecraft's stories often involved cults; they reflected Lovecraft's own deep-seated xenophobia, racism, and sexism; and, above all else, they have a fleshy horror to them, a deep revulsion at the idea of the human body as a biological machine. There is something of that horror regarding flesh in the views of the rationalists and Yudkowsky. Like Kurzweil, they want to live forever. All the rest of it seems to be a fantasy spun out of that desire and the overwhelming fear of death that comes with it. It's the flesh that dies; it's flesh that can't go to space. Flesh is simply not strong enough to support the fantasies of immortality and growth that the rationalists entertain. So flesh must be discarded, if the fantasy is to be maintained. The human body is the enemy.

Perhaps the rationalists' AI apocalypse is this overwhelming fear of personal death, projected outward. (When Yudkowsky was asked about this, he gave a familiar reply: "Psychologizing people is a poor substitute for debating the scientific questions. I'd retort that they're probably psychologizing me on account of feeling a deep sense of inadequacy with their own ability to grapple with scientific questions, which leads them to try to bring up personal psychology as a distraction from an argument they know on some level that they'll lose.")[182] In 2023, the podcaster and computer scientist Lex Fridman asked Yudkowsky what life advice he'd give to high school and college students. "Don't expect it to be a long life," Yudkowsky replied. "Don't put your happiness into the future. The future is probably not that long at this point. But none know the hour nor the day."[183] In his *Time* op-ed that same year, Yudkowsky advocated risking nuclear war to avoid the slightest chance of a misaligned AGI. When you're staring down an apocalypse that only you can see, it makes things simple—nothing else can possibly be as important. Yudkowsky has been deploying that same logic for years. In 2010, when asked for advice about how to save humanity, Yudkowsky said, "Find whatever you're best at; if that thing that you're best at is inventing new math of artificial intelligence, then come work for the Singularity Institute. If the thing that you're best at is investment banking, then work for Wall Street and transfer as much money as your mind and will permit to the Singularity Institute where [it] will be used by other people."[184]

This echoes Will MacAskill's "earn to give" philosophy, the logic he successfully pitched to Sam Bankman-Fried. The effective altruists and longtermists grew out of the same transhumanist milieu at roughly the same time; the journalist Tom Chivers calls the two movements "conjoined twins."[185] Bostrom exported Yudkowsky's arguments about AI alignment to professional philosophy; MacAskill and Ord's followers are frequently readers of LessWrong and *HPMOR*. The two groups share a culture. And the effective altruists have carried that culture with them as they have penetrated the halls of power outside of the Silicon Valley bubble. They have brought concerns about AI alignment to the halls of Oxford and Cambridge; they're advising heads of state in the United States and United Kingdom; they're taking over think tanks at the heart of the military-industrial complex. Part of their success is likely due to optics: the effective altruists seem less bizarre than the rationalists. But that's just an illusion. "These people make you feel like a conspiracy theorist. It sounds ridiculous," says Gebru. "Their whole deal is so ridiculous that people don't want to believe it."[186] But under the surface, the effective altruists are, if anything, even more bizarre than Yudkowsky's apocalyptic fever dreams.

4

THE ETHICIST AT THE END OF THE UNIVERSE

On the southwestern outskirts of central Oxford, across the railroad tracks from most of the rest of the university, sits a small three-story building. The building houses several academic research institutions, but it looks and feels more like a small tech start-up. It has all the trappings and perks of a San Francisco software company: abundant snacks, hot desking, and phone and video chat booths; a free cafeteria, totally vegan; a gym and showers on the ground floor; common areas and workspaces with open floor plans. A small picnic table sits on Astroturf out front. In back, the Thames ambles past the floor-to-ceiling cafeteria windows. The hallways are dotted with glass marker boards, covered in mathematical scrawls and notes about everything from Tolstoy to yesterday's lunch. One of them asks a question in off-white ink and large, friendly handwriting: *What would you do if all world problems were solved?*

- *Organise a weeklong festival for everyone to celebrate the end of suffering*, says one reply.
- *Reminisce about the close call we had with friends :-)* says another.
- *Sleep probably*
- *Work as a carpenter*
- *Explore altered states of consciousness*
- *Read the Tempest (presumably Shakespeare's last work and vision)*
- *Sit on the beach and build sandcastles all day*
- *Pretend they weren't solved, try (pretend to try?) to solve them again, and derive meaning from that.*

This building, Trajan House, is the academic home of the Centre for Effective Altruism. It's also the home of the Global Priorities Institute, Effective Ventures, the Centre for the Governance of AI, and the Future of Humanity Institute—all EA-affiliated institutions. Several dozen people work here, mostly graduate students and postdoctoral researchers, along with a smattering of permanent staff and faculty including William MacAskill, Toby Ord, Anders Sandberg, and Nick Bostrom.* It's May 2023, and I've traveled here from California to talk with MacAskill, but he's canceled at the last minute. Instead, I'm talking with Sandberg, a senior research fellow at FHI, with an office not far from MacAskill's. Sandberg has been a fixture in transhumanist and singularitarian communities for decades, going back to the days of the Extropian email listserv in the 1990s, where he first encountered a teenager by the name of Eliezer Yudkowsky. "I'm very much a skeptical transhumanist who loves being annoying to other transhumanists," he tells me. "I've been involved long enough that now I get a grumpy old man card."

* FHI and Bostrom are no longer there, but they were at the time of my visit.

Sandberg has an unassuming and jovial demeanor, and he is just as happy expounding on the distant future of humanity as he is talking about the view from his office window or the differences between Oxford and his native Sweden. He is stocky and blond, and he wears a thin medallion around his neck, a small disc of metal with a few lines of text pressed into it. He smiles when I ask him about it. "Oh yeah. This is for cryonics," he says. "I'm signed up to be frozen if I'm dying. And I'm kind of 'out of the freezer.' I'm wearing it openly because it's a brilliant conversation starter." I point out that cryonics isn't seen as mainstream science or medicine, and Sandberg agrees that the freezing process would burst some cell membranes throughout the body. "We need to solve those problems. The real way maybe to prove that you can't revive somebody is to say that some crucial information actually does get scrambled. That would be a way of actually disproving cryonics properly. I haven't seen anybody do a proper job of this." When it comes to handling the damage incurred by the freezing process, he continues, "the standard assumption [among cryonics advocates] these days tends to be, OK, let's let the nanomachines sort it all out." But, he says, "most normal medical people of course think that this is just crackpottery, pseudoscience, or a new version of Egyptian mummies. But then again, low temperature in surgery has become a big thing. So it's kind of interesting to see how they accept some parts of it, but not others."[1] (Yudkowsky used to advocate for cryonics too. He wrote in 2010 that "if you don't sign up your kids for cryonics then you are a *lousy parent*.")[2]

Yet there are good reasons cryonics isn't accepted among doctors and scientists. Nanorobots of the sort Sandberg is positing are purely hypothetical, and there are good reasons to think they couldn't exist. But that's not even the biggest problem with freezing dead or dying people in hopes of reanimating them. "There is absolutely no current way, no proven scientific way, to actually freeze a whole human down to that temperature without completely destroying—and I mean obliterating—the tissue," says Shannon Tessier, a cryobiologist at Harvard and an expert on suspended animation in animals. If you try to freeze human tissue, she continues, "the tissue is completely obliterated, the cell membrane is completely destroyed. So there's actually no proof that you're preserving anything, and that's because the science is just not there yet."[3] Michael Hendricks, a neurobiologist at McGill University, is even more blunt. "What's being done now in the commercial cryonics industry is garbage. They're making puddles of pink mush in a liquid nitrogen tank. It's nothing that could ever be used for anything," he tells me. While it might eventually be possible to chill living people to a temperature near freezing and keep them in a kind of hibernation state, Hendricks says that "when you get into people who have died, and then the cryonics people show up with their head saws—it's too late by the time you get in there with a saw. The tissue starts breaking down, and importantly, it's pretty fast."[4]

Sandberg is sanguine in the face of these concerns, suggesting that plastination—a method of preserving detailed anatomical structures by replacing large portions of the body with plastic—might keep enough information about the brain intact after death to allow the eventual scanning and uploading of the consciousness of a dead person, bringing them back to virtual life in the future. "Yeah, in the end, you could say, 'Oh, it's a matter of faith.' But it's a faith that you can justify to some degree with some evidence," he tells me.[5] But uploading a mind from a healthy living brain is probably impossible; uploading a mind from a frozen or plastic dead brain is even less likely to be possible. "Let's say we imagine some far future technology that can do these things that we don't know anything about now," Hendricks says. "What are the chances that we are preserving tissue in such a way now that it is going to be backward compatible with some far future technology? It's basically thinking you could stick a papyrus in your USB drive to read it. It's just not going to be the right thing."[6] Rather than starting the conversation, Sandberg's medallion has led to an end: I don't know what else to say in the face of his blithe unconcern with the scientific implausibility of the futures he's describing.

Down the hall from Sandberg is another major figure in effective altruism, the ethicist Toby Ord. Ord cofounded Giving What We Can and the CEA with MacAskill; Ord and MacAskill also jointly coined "longtermism" and helped choose the name "effective altruism." Ord's not around that day, but we talk a few weeks later over video chat. Compared to Sandberg, he comes across as less speculative and more conservative in his views about the future—and more worried about what that future might hold for us. "I think that this is a special time," he tells me. Specifically, he thinks that we live at a time when the risk of humanity going extinct is higher than it's ever been. Ord wrote a book on existential risks to humanity, called *The Precipice*—referring to the particularly risky time Ord sees us in—which came out a couple of years before MacAskill's *What We Owe the Future* and serves as a kind of prequel to it. In his book, Ord estimates the risk of existential catastrophe in the next century at around one in six, roughly 17 percent. That estimate includes more than just the extinction of humanity. Instead, it's his best estimate for the odds of the "destruction of our potential," as he puts it. "There could still be humans left, but in such a way that, for example, on a planet that's been environmentally destroyed, and they can no longer achieve anything close to what they could have achieved. That kind of idea."[7]

"My one in six [estimate] is meant to be pretty rough," he adds. "If you said, 'I think it's one in ten,' I would say, 'Wow, we agree!' If you said it's one in sixty, I would say, 'Oh, sounds like we actually have a disagreement.' And if you said it's one in a billion, I would say"—and at this point Ord breaks out laughing in imagined exasperation—"look, we clearly have different opinions on this. And we should try to get to the bottom of them." But why one in six, I ask? Where did he get that number? "I don't have a simple recipe for creating my number," he says. "I guess it would ultimately be like, live my experiences or something through my whole life." But, he explains, most of that one-in-six chance of catastrophe comes from his estimation of the risk from two particular threats to humanity's survival.

One of these two major threats—the smaller one, in Ord's estimation—is the risk of engineered pandemics. "I ended up putting [that] around 3 percent," he tells me. He's not talking about natural pandemics; he's talking about biological weapons. "In this case, the engineered pandemics would be being deliberately created in order to be dangerous."[8] But as terrifying as that prospect is, there's another that Ord thinks is more dangerous. Given his aforementioned concern about ecological disaster permanently crippling humanity, you might guess that the biggest risk Ord sees facing us right now is global warming. But he rates that as fairly low on the scale: he gives global warming only a one-in-a-thousand chance of causing the extinction of humanity or sending human civilization into unrecoverable collapse in the next century, the same odds he gives for nuclear war.[9] Instead, most of Ord's estimated odds come from just one speculative—yet familiar—risk. "A bunch of it's coming from AI," he tells me. The reasons he gives are broadly the same as the rationalists' logic—indeed, it was Bostrom who put existential risk on Ord's radar. A misaligned AGI, Ord says, could easily destroy humanity. "I ended up putting [the odds of that] at about 10 percent. And that 10 percent is bigger than the 3 percent" from engineered pandemics, Ord continues, partly because a misaligned AGI "would be an intelligent adversary, rather than something that's just biologically optimized to be successful, something that could try to outthink us and out-plan us."[10]

It's rather breathtaking to see an Oxford ethics professor state that the danger over the next century from an ill-defined hypothetical technology is fifty times greater than the danger posed by global warming and nuclear weapons combined. But Ord isn't even the only person matching that description who works in that building. In *What We Owe the Future*, MacAskill largely dismisses global warming as a serious threat to the survival of human civilization:

> Warming of seven to ten degrees [Celsius] would do enormous harm to countries in the tropics, with many poor agrarian countries being hit by severe heat stress and drought. Since these countries have contributed the least to climate change, this would be a colossal injustice. But it's hard to see how even this could lead directly to civilisational collapse. For example, one pressing concern about climate change is the effect it might have on agriculture. Although climate change would be bad for agriculture in the tropics, there is scope for adaptation, temperate regions would not be as badly damaged, and frozen land would be freed up at higher latitudes. There is a similar picture for heat stress. Outdoor labour would become increasingly difficult in the tropics because of heat stress, which would be disastrous for hotter and poorer countries with limited adaptive capacity. But richer countries would be able to adapt, and temperate regions would emerge relatively unscathed.[11]

He goes on to consider more unlikely scenarios, and (hesitantly) dismisses them as well, stating that his "best guess is that global agriculture would still be possible. . . . Even with fifteen degrees of warming, the heat would not pass lethal limits for crops in most regions."[12]

Speaking on an effective altruist podcast in 2020, Ord made a similar but even more extreme claim. "There has been some analysis of, if you had very large amounts of warming, such as 10 degrees [Celsius] of warming, would it start to make areas of the world uninhabitable?

And it looks like the answer is yes," he said. But, he added, "it really just suggests that the habitable part of the world would be smaller. . . . And it seems hard for me to think that, given it wouldn't be that much smaller, as to why then civilization would be impossible, or a flourishing future would be impossible in such a world. That just doesn't seem to have much to back it up at all." Ord ultimately concludes that human civilization has a good chance to survive even at double that temperature rise. "I looked at these models up to about 20 degrees of warming, and it still seems like there would be substantial habitable areas," he said. "But, it's something where it'd be very bad, just to be clear to the audience," Ord hastened to add.[13]

Climate science suggests that "very bad" is a gross understatement. "A temperature rise of 10 degrees [Celsius] would be a mass extinction event in the long term," says Luke Kemp, a researcher at the University of Cambridge and an expert on climate-induced civilizational collapse. "It would be geologically unprecedented in speed. It would mean billions of people facing sustained lethal heat conditions, the likely displacement of billions, the Antarctic becoming virtually ice-free, surges in disease, and a plethora of cascading impacts. Confidently asserting that this would not result in collapse because agriculture is still possible in some parts of the world is silly and simplistic."[14] Andrew Watson, a climate scientist at the University of Exeter, was one of the experts MacAskill's team of assistants contacted as part of the background research for *What We Owe the Future*. "I completely disagreed with their statement that warming of 7 to 10 degrees and more would not seriously endanger civilization," he tells me. "I agree with their statements about the harm that would be done, I also agree that, most probably, humans would survive, but I think it very likely that our current civilization would collapse under these extreme warmings, and that the size of the human population would be greatly diminished."[15]

It's not that MacAskill and Ord don't take climate change seriously—they clearly do. In his book, MacAskill repeatedly mentions that humanity has to stop using fossil fuels. On that EA podcast, Ord said that he thinks that climate change is dangerous, "much worse than if it wasn't happening," and we need to do what we can to stop it. The problem is that MacAskill and Ord don't take climate change seriously *enough*. "While there are a lot of things which could very clearly cause a very large amount of human misery and damage, it's quite unclear how [global warming] could cause the extinction of humanity or some kind of irrevocable collapse of civilization," Ord said on the podcast. There are two issues here. The first is that Ord and MacAskill are out of step with the scientific mainstream opinion on the civilizational impacts of extreme climate change. In part, this seems to stem from a failure to imagine how global warming can interact with other risks (itself a wider issue with their program), but it's also a failure to listen to experts on the subject, even ones they contact themselves.[16] "When I try to imagine a world in which the tropics are largely uninhabitable and unproductive (from 30 degrees north to 30 degrees south, that is half the area of the planet and most of the biological productivity) I find it hard to believe that there won't be such tipping points due to famine, migration and conflict, that will make it difficult for our civilization to survive," Watson tells me.[17]

Science can help fill in that imaginative gap: the last time global temperatures were 15 to 20 degrees Celsius higher than they are now was about fifty-six million years ago, during the Paleocene-Eocene Thermal Maximum (PETM). Back then, there were no polar ice caps at all. Deep in the Arctic Circle, an island that would one day be part of far northern Canada was home to a swamp forest, filled with palm trees, giant tortoises, alligators, and flying lemurs. The Arctic waters could have reached a balmy 24°C (75°F); in the tropics, the ocean was 36°C (97°F), or perhaps even warmer. Average summer surface temperatures on land were close to human body temperature (or slightly higher) at the latitude of Portland or Beijing. But Portland and Beijing would both have been underwater: sea level was 50 to 200 meters (160 to 660 feet) higher than it is today.[18] This is manifestly not a world that advanced human civilization could survive in.

Mercifully, warming of that magnitude is not considered likely by most climate models—unless humanity doesn't make any effort to reduce our collective carbon emissions into the atmosphere, in which case global temperatures might get close to this range three hundred years from now.[19] The tipping points that Watson described are certainly within easy reach if we choose to do nothing. And Watson himself says that even worse outcomes than a return to the hothouse of the PETM climate can't be ruled out. If humanity continues to burn gas, oil, and coal, Watson estimates a one-in-a-hundred chance that Earth enters a "runaway greenhouse" effect, boiling off the oceans and leaving Earth like our neighboring world Venus—a planet shrouded in an insulating atmosphere, where the surface is hot enough to melt lead. Given that, Ord and MacAskill's confidence that climate change probably doesn't pose the kind of existential threat they're worried about is unwarranted. And the fact that they're primarily worried about existential threats in the first place is the other problem: once a threat has been deemed existential, it's impossible to outweigh it with any less-than-existential threat in the present day. In determining which threats count as "existential," much ends up depending on estimations of risk, estimations that—apparently—aren't informed by the best available science and expert opinions. "MacAskill falls into the trap of being 'foolishly wise,'" says Watson. "Wise fools [are] those who recognise the limits of their knowledge and therefore are able to take some account of what they don't know. I think of the opposite as the foolish wise, very clever people whose erudition and wide book-learning lead them however to inadequately account for the 'unknown unknowns.'"[20]

Despite these important mistakes at the core of their worldview, Ord, MacAskill, Sandberg, Bostrom, and the rest of those who work at Trajan House aren't fringe. The building itself is testimony to how influential their views are and how successful their movement has been. It screams of money, from the corporate office furniture to the snack drawers filled with cruelty-free chocolate bars.[21] Financial records show a little over $75 million in donations to Effective Ventures in their 2021 fiscal year alone, most of which came from Open Philanthropy.[22] In fact, almost a third of CEA's entire income that year came from a pair of Open Philanthropy grants for buying and renovating an Oxfordshire manor house, Wytham Abbey, for use as an EA event space.[23] This led to some spirited debate among the rank-and-file effective altruists about the justification for spending that much on a venue, rather than putting the money toward, say, malaria nets.[24] But effective altruists dismayed by the purchase ultimately got what they wanted, in a way: after returning funds to FTX's creditors in the wake of Bankman-Fried's fraud conviction, Effective Ventures decided to put Wytham Abbey up for sale in 2024.[25]

Yet SBF isn't even the most notorious billionaire directly connected to the leaders of the EA movement. On March 29, 2022, about two weeks before Elon Musk publicly announced his offer to buy Twitter, MacAskill texted Musk. (MacAskill got Musk's number through his friend Igor Kurganov, an effective altruist and associate of Musk's.)[26] MacAskill offered to introduce Musk to Sam Bankman-Fried in order to help with purchasing Twitter—something that, according to MacAskill, SBF had been contemplating himself. "Does he have huge amounts of money?" Musk asked in reply. "Depends on how you define 'huge'! He's worth $24B," MacAskill replied. "I asked and he said he'd be down to meet you." "You vouch for him?" asked Musk. MacAskill replied, "Very much so! Very dedicated to making the long-term future of humanity go well." MacAskill made the introduction and the three men texted briefly, though there are conflicting reports on whether anything came of it. Once again, there were members of the EA community asking why their leadership was involved with such a purchase at all.[27]

Effective altruists are also using Silicon Valley money to enter the halls of political power. The Center for Security and Emerging Technology (CSET) at Georgetown University is a think tank that "provides decision-makers with data-driven analysis on the security implications of emerging technologies," according to their own website. "CSET is currently focusing on the effects of progress in artificial intelligence (AI), advanced computing and biotechnology."[28] CSET also boasts of their broad and deep connections within the US government and prestigious institutions around the country. "Former CSET staff are now working for the National Security Council, the Office of Science and Technology Policy, Congress and other parts of the U.S. government, as well as Stanford University," said a 2021 press release. "Current staff are serving as fellows in both the legislative and executive branches [and] CSET research has been cited in reports by the White House."[29] This level of impact is especially surprising given how new CSET is. It was founded in 2019 with a $55 million grant from Open Philanthropy; later grants raised that organization's total support for CSET to about $100 million as of 2023.[30]

Open Philanthropy has funded myriad EA organizations, but its spending on CSET puts that think tank into a different league. Upon its founding, CSET was immediately a major player. "We're the largest center in the U.S. focused on AI and policy," said Jason Matheny, the founding director of CSET, in 2019.[31] Matheny's involvement was a large part of the reason Open Philanthropy was willing to make such a large grant for CSET in the first place.[32] Matheny had extensive experience in government and deep ties to the EA community. Prior to taking the job at CSET, Matheny had been assistant director of National

Intelligence and director of the Intelligence Advanced Research Projects Activity, as well as a member of the National Security Commission on AI—and a few years before that, he'd been a research associate at FHI.[33] He wrote a paper in 2007 about existential threats to humanity such as misaligned AGI and cited Bostrom's work approvingly.[34] He is also a longtime friend of Toby Ord.[35] "There are some decisions that are made only by governments, and some of those decisions are highly consequential," Matheny said in a talk at EA Global, the flagship EA conference, in 2017. "They include decisions like going to war, or what weapon systems will be fielded, or how technologies will be embedded within larger critical systems. It makes sense to engage more effective altruists within these positions where they can influence those decisions."[36]

But today, Matheny isn't the head of CSET anymore. He's moved on to bigger things. In 2022, he became the president and CEO of the RAND Corporation, among the biggest and most influential think tanks in the world, especially regarding technology and military policy.[37] Under Matheny's leadership, RAND's influence on US AI policy has grown. "Many key personnel at top AI companies are advocates of effective altruism," wrote Brendan Bordelon in *Politico* in late 2023. "Now RAND, an influential, decades-old think tank, is serving as a powerful vehicle through which those ideas are entering American policy."[38] Meanwhile, Open Philanthropy has started giving out hefty grants to the RAND Corporation under Matheny's leadership, to the tune of $26 million in 2023 alone, including $10.5 million for "potential risks from advanced artificial intelligence."[39] Open Philanthropy has also been funding a fleet of policy fellows across the federal government through another nonprofit, the Horizon Institute for Public Service. "A sprawling network spread across Congress, federal agencies and think tanks is pushing policymakers to put AI apocalypse at the top of the agenda—potentially boxing out other worries," Bordelon wrote. "Current and former Horizon AI fellows with salaries funded by Open Philanthropy are now working at the Department of Defense, the Department of Homeland Security and the State Department, as well as in the House Science Committee and Senate Commerce Committee, two crucial bodies in the development of AI rules."[40]

The effective altruists have even tried to get one of their own elected to Congress. Carrick Flynn—a former researcher with FHI, CSET, and the Centre for the Governance of AI—ran in the Democratic primary for an open House seat in the new Sixth District of Oregon in 2022, a competitive race in a swing district.[41] Flynn, born in the district but a newcomer to politics who had only voted twice before, pitched himself as an expert on pandemic preparedness.[42] It's true that Flynn had worked on pandemics during his time at various EA-affiliated institutions, but that was hardly Flynn's exclusive area of work: while he was at FHI, he was also one of the coauthors on a paper about economic policy in a future where a tame superintelligent AGI has enabled a few companies, or a single company, to capture much of the money in the world.[43] (This sounds similar to Sam Altman's ideas, and that may not be a coincidence: two of the other authors on that paper ended up working for Altman.) For a time, it seemed that Flynn's lack of political experience and research on fringe AI scenarios was not a serious impediment to his standing in the race—not with over $10 million in support for his campaign from a political action committee funded almost entirely by Sam Bankman-Fried. That included nearly a million dollars of attack ads against Flynn's main opponent in the primary, Andrea Salinas, a state legislator.[44] Polls showed a close race in the primary, but ultimately Flynn came in a distant second to Salinas, who went on to win a genuinely close race in the general election that fall.[45]

Despite all that, EA's influence at the highest levels of politics is arguably stronger in the United Kingdom than in the United States.

Dominic Cummings, an advocate of effective altruism, was cofounder and head of the 2016 Vote Leave campaign in the UK, supporting Brexit.[46] When Boris Johnson, another Brexiteer, became prime minister in 2019, he appointed Cummings as his chief adviser and de facto chief of staff.[47] "We're hiring data scientists, project managers, policy experts, [and] assorted weirdos," Cummings wrote on his blog several months later. He opened the post with a quote from Eliezer Yudkowsky about inefficiency in government; his blogroll links to MIRI, Yudkowsky, and Siskind. "We want to hire an unusual set of people with different skills and backgrounds to work in Downing Street," he wrote in the post, including "weirdos from William Gibson novels."[48]

Cummings was dismissed from his post in November 2020; he later claimed that Johnson was "unfit for the job."[49] Johnson himself resigned in September 2022, but that was not at all the end of EA's currency at the highest levels of the UK government.[50] Three days before I met with Sandberg at Trajan House, I'd landed in London, my jet-lagged brain leaking out of my ears after a ten-hour flight across eight time zones. I'd been thinking about AI and the future of technology the entire ride over, between unsuccessful attempts to sleep on the plane. Now, I just wanted to get my bearings and stay awake until a reasonable hour. As I stumbled through the airport, I looked at the *Guardian*, thinking that I should know something about recent UK news if I was going to be there for the next few days—and that it would be good to give my brain a break from claims of AI doom. It was a nice idea while it lasted. "Rishi Sunak Races to Tighten Rules for Artificial Intelligence amid Fears of Existential Risk," blared the top headline on the *Guardian*'s website that afternoon.[51] "[Prime Minister] Rishi Sunak is scrambling to update the government's approach to regulating artificial intelligence, amid warnings that the industry poses an existential risk to humanity," the article said. "Last week, Sunak met four of the world's most senior executives in the AI industry, including Sundar

Pichai, the chief executive of Google, and Sam Altman, the chief executive of ChatGPT's parent company OpenAI. After the meeting that included Altman, Downing Street acknowledged for the first time the 'existential risks' now being faced."[52] Sunak's government was in the midst of setting up an AI safety task force, with major input from effective altruists at all levels of UK AI policy.[53]

Meanwhile, the day after I visited Trajan House, a text from a friend pointed me to more AI safety news. "Today many of the key people in AI came together to make a one-sentence statement on AI risk," Ord tweeted that day.[54] "Mitigating the risk of extinction from AI should be a global priority alongside other societal-scale risks such as pandemics and nuclear war," the statement read.[55] The statement and its release was organized by the Center for AI Safety, yet another nonprofit funded by Open Philanthropy—and by a $6.5 million grant from the FTX Foundation, according to court documents in the FTX bankruptcy case.[56] The statement was intended to make a splash, and it did. It was reported on by the *New York Times*, the BBC, the Associated Press, NPR, and the *Washington Post*, among many others.[57] The statement was signed by nearly seven hundred tech executives, AI scientists, and other assorted academics and luminaries, including Sam Altman, Bill Gates, Peter Singer, Dustin Moskovitz, Ray Kurzweil, David Chalmers, Eliezer Yudkowsky, and the musician Grimes. Ord signed too, as did a notable former member of the FTX Foundation team: one William MacAskill.[58]

* * *

"I was pretty alarmed by Bing. . . . I was not expecting 2023 to be the year where an AI chatbot was released by a major company, which threatened to kill an AI ethics researcher and also threatened revenge upon a number of people—threatened revenge on an Associated Press journalist for writing a negative story about it, things like that, to expose him for crimes—without any apology coming from the company," Ord says to me. I'd asked him if there were any existing computer programs that really scared him. "[Bing] didn't exactly scare me," he tells me. "I don't think that it can enact revenge and so on. But it still was remarkable that such a thing got released and didn't get picked up in checks." But, he adds, LLMs have a lot of problems too. "If you ask [GPT-4], 'Can you stack a cube, a cylinder and a square-based pyramid?' it can give you faulty answers to that question, and say, 'Yeah, first you put the square-based pyramid down on its square face, then you put the cube on top of one of the triangular faces, and then you put the cylinder on top of the cube.' And so it's highly uneven in its capabilities."[59]

It's a little confusing to hear Ord say that he finds threats issued by LLMs alarming when he seems to understand their limitations. An LLM saying it wants to kill you shouldn't be more troubling than an LLM saying it knows the sky at night is neon pink; they both come from the same place. All they can do is hallucinate. But, Ord explains, it wasn't so much the Bing chatbot itself that concerned him. He sees it as a symptom of a cavalier attitude toward AI safety at Microsoft. "I was mainly alarmed that it was released, to be honest." Ord claims his fears about insufficient guardrails on superintelligent AGI are shared by many experts in AI research. "It turns out that the average ML researcher thinks that there's something like a 5 percent chance of" superintelligent AGI murdering everyone, he tells me. "So it's not that I'm out on a limb or something, either."[60]

There are surveys that suggest Ord is right, which he pointed me to when we spoke. Most notably, the results of a 2022 survey led to the claim that "half of AI researchers believe there is a 10% chance or greater [that] human extinction will result from AI."[61] But this survey has serious problems: it may not have actually surveyed a representative cross-section of the field, and it hasn't gone through peer review.

"What happens is that somebody surveys people who have published at particular conferences in AI and machine learning, and they send it out to thousands of people, they get on the order of several hundred responses, and they asked people to assign probabilities," Melanie Mitchell tells me. "Who are these people? What is their expertise? How do they come up with these probabilities? Are people just kind of guessing, with no basis? Do they have any confidence? . . . I think these surveys are not super useful. And the history of AI is littered with failed predictions."[62]

Part of the problem, according to Mitchell, is that there isn't even consensus in the field about what AGI is in the first place. "The goal is not well defined, this goal of AGI. What does it even mean? Does it mean the same thing to different people who are assigning probabilities? If you look at the actual data on these surveys, people are just all over the place . . . which means that nobody really knows." Without that definition in place—and without any idea of what it would actually take to build an AGI—these surveys lead to a false sense of accuracy about a fundamentally undefined question. "I remember reading some article that described . . . someone as an AGI expert," Mitchell tells me. "What does that even mean? What's an AGI expert? There are no AGI experts. So, you know, I don't think people really know what it is they're predicting, and what the probabilities even mean." Mitchell went on to point out a crucial difference between concerns about existential risk from AI and evidence-based concerns about catastrophes like global warming. "Climate science has a huge consensus," she says. "I don't think we have any evidence or consensus on a lot of the AI risks discussion. There are certainly AI risks that are happening right now that we certainly have a lot of evidence for, like facial recognition having biases and other things. But for things like so-called existential risk, it's really based on speculation—not experiments, not evidence."[63]

Given the thinness of the evidence for an impending AI apocalypse, how can Ord be sure of the odds he's assigning to it? He's not. "I think a bunch of this comes down to what happens when you don't know, what happens when there's disagreement or uncertainty," he tells me. As for his estimate of the risk from unaligned AGI and other imagined existential threats, Ord says that he's not trying to be strict or prescriptive about it. When it comes to his one-in-six estimate of extinction or collapse in the next century, he says, "it's more of attempting to communicate Toby Ord's credences about this probability, rather than trying to say, 'Here's a credence that you should have. And that if you don't have it, and if you don't get it, then you're making an error.' It's not meant to be that."[64]

But for all that Ord is modest about his probability estimates, they do matter, and not just to Ord. He is a leader in the effective altruist community; his words carry weight. And by longtermists' logic, probability estimates like Ord's can be enormously important when trying to decide what actions to take, here and now, to improve the world. The typical longtermist game involves multiplying a very small number by a very large number in order to estimate how many people you might help in the future by doing a particular thing now. One of the problems with that game is that it's extremely sensitive to small changes in the numbers being used—changes so small that they're basically imperceptible, and very difficult to reason about.

Take the example from Chapter 1, where there's a one-in-10^{17} chance of helping a cosmically huge future with 10^{40} people in it. That led to a situation where you should take that small chance over an action that is certain to help every person alive today, because that tiny chance, multiplied by such a large number of people, would result in a number of people being helped that is far larger than the current population of Earth. But say that the odds of helping everyone in that future were lower than you originally thought—say they were one in 10^{39}.

Those are still really low odds. In fact, they're so low that it's hard to say that you could reliably tell the difference between a one-in-10^{17} chance and a one-in-10^{39} chance. They're both so close to "never, ever going to happen" that the difference seems impossible to discern, especially when reasoning about something as fuzzy as the impact of a charitable donation in the far future. (Enter a lottery with the former odds once a second, and it'll take around three billion years for you to win, on average. If the lottery has the latter odds, it'll take ten billion trillion times longer than that. The second game certainly has much worse odds than the first, but after the first million years of constant disappointment, would it really make a difference to you?)

Yet this difference does matter for the longtermist argument: If the odds of your donation meaningfully helping that astronomically huge number of people in the future were really one in 10^{39}, then suddenly it's a bad bet. You'd only be helping ten future people, on average. You'd be doing much more good by helping everyone alive today. So the whole line of reasoning depends on your ability to tell the difference between one really small probability and another even smaller one, all in the context of forecasting the effects of your actions trillions of years or more into the future, across the entire observable universe. And the large numbers used in these arguments are uncertain in a similar way. If the number of future humans is 10^{20} in the scenario from Chapter 1, that also suggests that the right move is to help everyone alive today. Population estimates one hundred years out for humans on a single planet are already difficult to perform. How confident can you really be in forecasting the number of humans that will live from now until the end of time?

Longtermists often make arguments depending on such forecasts, and they often invoke speculative future technologies while doing so. In a 2012 paper, Nick Bostrom estimated that, using brain uploading and simulations, the equivalent of 10^{52} human lives "of ordinary length" could be lived out in computers across space over the future history of the universe. If there's even a 1 percent chance that such a future will happen, he wrote, then "the expected value of reducing existential risk by a mere one billionth of one billionth of one percentage point is worth a hundred billion times as much as a billion human lives."[65] MacAskill was even more specific in a paper he wrote in 2021 with Hilary Greaves titled "The Case for Strong Longtermism." Discussing the cost-effectiveness of funding work on "AI safety" (that is, preventing a superintelligent AI from destroying humanity), they cite the same surveys that Ord pointed to as justification for their estimate of the existential risk from AI, and then claim that "even a highly conservative assessment would assign at least a 0.1% chance to an AI-driven catastrophe (as bad or worse than human extinction) over the coming century." Combining that estimate with a guess about the efficacy of investment in AI safety research and with their "reasonable estimate" of the number of future beings—which they claim is "at least" 10^{24}—MacAskill and Greaves arrive at a stunning conclusion. "Every $100 spent [on AI safety] has, on average, an impact as valuable as saving one trillion [lives] . . . far more than the near-future benefits of [malaria] bednet distribution."[66] For a strong longtermist, investing in a Silicon Valley AI safety company is a more worthwhile humanitarian endeavor than saving lives in the tropics.

This is not an isolated problem; it's been part of longtermism from the start. Nick Beckstead, an AI safety consultant, has a long history with longtermism and EA. He was CEO of the FTX Future Fund before FTX imploded in 2022. Before that, he did a stint at FHI in Oxford, followed by a job at Open Philanthropy.[67] And before all that, he did a PhD in philosophy.[68] Beckstead's 2013 PhD thesis, titled "On the Overwhelming Importance of Shaping the Far Future," is cited by Ord as one of the major foundational works on longtermism and existential risk.[69] In that thesis, Beckstead says the following:

> To take another example, saving lives in poor countries may have significantly smaller ripple effects than saving and improving lives in rich countries. Why? Richer countries have substantially more innovation, and their workers are much more economically productive. By ordinary standards—at least by ordinary enlightened humanitarian standards—saving and improving lives in rich countries is about equally as important as saving and improving lives in poor countries, provided lives are improved by roughly comparable amounts. But it now seems more plausible to me that saving a life in a rich country is substantially more important than saving a life in a poor country, other things being equal.[70]

In other words: the lives of people living in, say, Mozambique matter less than the lives of people living in the United States, according to Beckstead, because the people in the United States will contribute more to the glorious longtermist future in space.

Like Bostrom's apology for his racist email, Beckstead issued a statement about this portion of his thesis years later—and like Bostrom, Beckstead missed the point, but without even attempting to give an apology. He didn't repudiate his earlier view either; he merely attempted to clarify what he had meant. "This passage was exploring a particular narrow philosophical consideration, in an academic spirit of considering ideas from unusual angles," he wrote in 2022. "I do not believe that lives in rich countries are intrinsically more valuable than lives in poor countries."[71] This doesn't contradict or deny Beckstead's earlier statement, though: he may not think that the life of a person living in the United States is *intrinsically* more valuable than the life of someone who lives in Mozambique, but he can still think that, as a *practical* matter, a US life matters more—and indeed, that's exactly what he's suggesting in that passage of his thesis.

This isn't to say that Beckstead thinks it's OK to ignore the residents of poor countries. "All things considered, I believe that it is generally best for public health donations to prioritize worse-off countries," he wrote, "and I've personally focused significant amounts of my career on promoting such donations, e.g., as a founding board member of [Giving What We Can]." Beckstead concluded his statement with a request: "If you quote this part of my dissertation, I would appreciate it if you would also include this [note] to avoid unnecessary misunderstandings."[72] But such "misunderstandings" about what longtermism suggests regarding present-day poverty and inequality are rather common within the effective altruist community. "In the beginning, EA was mostly about fighting global poverty. Now it's becoming more and more about funding computer science research to forestall an artificial intelligence–provoked apocalypse," wrote Dylan Matthews, a journalist and effective altruist, while reporting on a major EA meeting in 2015. "Compared to that, multiple attendees said, global poverty is a 'rounding error.'"[73]

* * *

At lunchtime in Trajan House, I met up with David Thorstad, who was a research fellow there at the time. Working there, he found himself in the unexpected role of a kind of in-house critic of effective altruism. He hastened to point out that the effective altruists were very welcoming of thoughtful arguments criticizing their viewpoints, as long as they weren't presented in a polarizing way. "If you say, 'Effective altruists are nincompoops,' they'll be mad," he says. "But if you say, 'Effective altruists are worse than murderers because'—and to be clear, that's not my view—as long as you give them a 'because,' they'll listen." Thorstad also said there were patterns he'd noticed in the kinds of arguments that longtermists make. "They have one really good move, which is uncertainty," he told me over lunch. "So they'll often very quickly go to probability. They'll say, 'Well, won't you grant me that there's a one-in-a-million chance?' And you have to very much press [them], no, let's not jump straight into naming probabilities. Let's make arguments to ground our probabilities. And you really have to force people to give arguments, and it's often, I think, very hard to pull an argument out of people here."[74]

Longtermists, then, are making arguments with incredibly strong conclusions—funding AI safety research is trillions of times more cost-effective than preventing the spread of malaria! Saving a billion people today isn't as good as a minuscule chance of saving 10^{52} people who might exist someday!—based on arguments that rely on very small probabilities and that fall apart if those probabilities are wrong. And their estimates of those crucial probabilities are based on very little. Weigh that against the overwhelming evidence that there are people alive today who are in need, and the whole idea of longtermism looks shaky.

MacAskill and Greaves do mention in their paper on strong longtermism that these kinds of small-number-multiplied-by-big-number arguments are questionable, but they (tentatively!) conclude that such forms of moral reasoning are correct. Ord, for his part, isn't as sure. "I'm aware of arguments that involve very small probabilities and very large stakes, and I tried in *The Precipice* to not ever make that argument, and I think that I succeeded," he tells me. "In fact, I think I actively call out that argument and suggested we shouldn't make it." When you make arguments like that, he continues, "everything's really sensitive to these uncertain numbers. And that alarms me."[75]

But putting aside whether Ord actually succeeded in avoiding arguments of that form—and also putting aside his deeply questionable odds for AI catastrophe and climate disaster—he's still relying on another tenet of longtermist thought, as are MacAskill and Bostrom: the "time of perils" hypothesis. This is the idea that we are living through a uniquely important, dangerous, and temporary period in history—that we can destroy ourselves, but if we can just make it through the next few centuries, the risk will drop and humanity will be safe for thousands, millions, or billions of years. "Humanity has had rapidly escalating power," Ord says to me. "As of the development of atomic weapons, we've got to the stage where we arguably have the power to destroy ourselves through our own action. We'd always been vulnerable to risks from things like asteroids that could destroy us. But now, we have the possibility of taking our own lives. . . . We've had our power outstripping our wisdom, as Carl Sagan and others have noted."[76] The term "time of perils" also comes from Sagan's writings, and Ord borrows another analogy from Sagan for talking about this: the idea of humanity as a teenager, reckless and testing the limits of its newfound power.[77] "If we play our cards right, humanity is at an early stage of life: still in our adolescence; looking forward to a remarkable adulthood," Ord writes in *The Precipice*. "Like an adolescent, we are rapidly coming into our full power and are impatient to flex our muscles, to try out every new capability the moment we acquire it. We show little regard for our future."[78]

Ord, like all longtermists, thinks that future could be very long and very large, containing those vast numbers of humans that Bostrom, MacAskill, and others talk about. And that's why he needs the time of perils to be short: otherwise, a future like that just isn't very likely. "The time of perils hypothesis is a very strong claim," Thorstad tells me. "It says now it's very dangerous; that's a very strong claim. It says things will get much, much, much less dangerous; that's another strong claim. And it says they're going to stay that way, for a long time." Borrowing Ord's analogy, Thorstad continues, "If we grow into adulthood as a species we'll become mature enough to handle risks. And if we don't grow into adulthood as a species, we will never end the time of perils and we'll go extinct. So the thought is either we make it out of these couple of centuries, or we don't."[79] But, he tells me, there are a few big problems with that, chief among them the fact that, once the time of perils is past, the risk would have to become low and stay low—very low. "And it's going to stay [low] literally forever. . . . When you help yourself to a billion-year [future] for humanity, you have to do that. Because if you have risk high now, but we're going to survive for a billion years, you have to have risk dropping low very soon." And, Thorstad concludes, there's essentially no evidence to support that.[80] For humans to survive for a billion years, the annual average risk of our extinction needs to be no higher than one in a billion. That just doesn't seem plausible—and it seems even less plausible that we could know something like that this far in advance.

Even forecasting "only" a few centuries into the future is nearly impossible. "Here's why I'm underwhelmed by the case for long-termism. . . . The Seven Years' War is about as far in the past as 2300 is in the future. And the Seven Years' War had a causal impact on just about every country on the planet, in many cases a massive impact," the philosopher Brian Weatherson, a professor at the University of Michigan, writes. "But did it make those countries better or worse, richer or poorer, more or less just, etc? Who knows! The [what-ifs] are too hard, even knowing how one particular run of history turned out. Our ability to know what will change extinction likelihoods [in] 250+ years, and the size and direction of those changes, is worse than our ability to know the size and direction of the causal impact of past events. And we don't know that."[81]*

---

* Relatedly, longtermists seem to have difficulty forecasting relevant events even on much shorter timescales, as economist Tyler Cowen pointed out in late 2022. "Hardly anyone associated with [FTX] Future Fund saw the existential risk to . . . Future Fund, even though they were as close to it as one could possibly be. I am thus skeptical about their ability to predict existential risk more generally, and for systems that are far more complex and also far more distant. And, it turns out, many of the real sources of existential risk boil down to hubris and human frailty and imperfections (the humanities remain underrated)."

Despite the difficulty of tracing the impact of actions over the course of centuries, much less millions of millennia, Ord's favored solution to permanently lowering existential risk is to build social and political momentum toward mitigating such risks. "Say, for example, on the level of environmentalism, or things like that, in the way that that was a fairly rapid change in human beliefs from 1950 to 1970. Where, all of a sudden, this thing that was not really considered part of moral thinking was changed to be a fairly core part of moral thinking," he tells me. "Maybe we could do that again, and take seriously the continuation of humanity as a core constraint on our behavior. And then govern that with treaties, and so forth, effectively a constitution for humanity going forward to live within a kind of sustainable risk budget."[82] But if the threat of human extinction is high now, it's difficult to imagine how a series of treaties, or even a world government, could keep that threat low over billions of years. Even keeping it low over decades could introduce other problems, as Thorstad points out. "When you say we're going to become wiser as a species and therefore we're going to manage risk, we need to say quite precisely, what does it mean to become wise? Why do we think it's going to happen? And why do we think it's going to drive risk down?" Then Thorstad gets specific about one possible future:

> The devil is just in the details of saying how that'll happen. . . . For example, if you ask what we would do about bioterrorism, which is usually cited as the second highest risk behind artificial intelligence, what we would do is institute systems of surveillance and control. We would amplify powers of surveillance, powers of punishment, powers of attention, powers of very

Cowen, "A Simple Point About Existential Risk," https://marginalrevolution.com/marginalrevolution/2022/11/a-simple-point-about-existential-risk.html.

> closely monitoring the individual use of genetic sequencing and other technologies. And if we were to do this well enough to chop a factor of one thousand, or ten thousand, or one hundred thousand off of the threat, despite rapid, rapid, rapid increases in technology, we might need to take the world to more of a frightening place that most people would not only not count as a wise place, but maybe not count as an improvement.[83]

Thorstad also says that many effective altruists disagree with Ord. Instead of relying on human agreements like treaties and government regulations, they envision a different means of permanently dropping the probability of human extinction. This is linked to one of the reasons it's hard to forecast the likelihood of human extinction far into the future: as time goes on, new existential threats to humanity can arise. If nothing else, disasters that are highly unlikely on a per-century basis, like large asteroid impacts or super-volcanoes, become inevitable once you start looking at tens or hundreds of millions of years at a time. And new technologies, like nuclear power and genetic engineering, have brought with them new threats, like nuclear war and engineered bioweapons. This last fact is one that the longtermists (and rationalists) certainly understand; that's how they see the prospect of AGI. But, Thorstad says, many of them also see aligned AGI as the real solution to getting existential risk low and keeping it low. "What many effective altruists will tell you about the time of perils is the following," he tells me. "AI is going to get very, very smart very fast. It might kill us, but if it doesn't, it's going to be so smart it can foresee and prevent all future risks. That's the orthodox story." Seeing the bemusement on my face, he adds, "I'm not kidding."[84]

Thorstad is skeptical of this claim, to put it mildly. "The claim that humanity will soon develop superhuman artificial agents is controversial enough," he writes. "The follow-up claim that superintelligent artificial systems will be so insightful that they can foresee and prevent nearly every future risk is, to most outside observers, gag-inducingly counterintuitive."[85] Asked why the effective altruists believe this, Thorstad is at a loss. "I got nothing. I got nothing," he says to me. He adds that, despite how common this belief appears to be among effective altruists, there's only one place where the belief is even written down in any detail—a brief comment on the EA forum, used as a standard reference for the topic despite its lack of detail—and he suspects this is because the view is so difficult to actually defend.[86] "I have a sense that, increasingly, many arguments by effective altruists back up against very, very strong and very speculative claims about AI that most readers are not in a position to accept or take to be supported by evidence," he tells me. "And that we will reach a point with these claims where we have so little in common that it's sometimes hard to have an argument."[87]

This promise of a benevolent godhead, a superintelligent AI that foresees and solves all human problems, is the same goal that the singularitarians and the rationalists have: the reduction of all problems to judicious application of computer science. More broadly, it's the dream of technology as salvation from all threats. Technology doesn't solve social and political problems, any more than it causes them. The prospect of nuclear war was made possible through technology, but it's a live concern because of geopolitics. Humans could come together and choose to rid the world of nuclear weapons, just as we could come together to end global warming. Applying more intelligence and technology to these problems won't solve them; they're fundamentally political. There are myriad human problems that could spiral out into war or ecological disaster or some other unpredictable threat, none of which have technological fixes. The solution to, say, border disputes between India and Pakistan isn't throwing more technology at the problem. The idea that an aligned AGI (if that idea makes sense at all) would eliminate all existential threats just doesn't seem to have any basis in reality. Insofar as there's scientific evidence on the subject, it cuts hard against this conclusion. All the computing power in the universe won't help you predict the weather two months from now, much less forestall a conflict arising from complex human social dynamics.

In a sense, then, Ord is right. The major problems facing humanity right now are social, and social movements to address them are a great idea. He's just misidentified which kinds of problems are the most pressing: he thinks risk from AI outweighs global warming and everything else combined. And even he isn't immune to the promise of the standard transhumanist technological utopia, with enormous numbers of humans living in space. "We once thought ourselves limited to the Earth; we know now that vastly greater opportunities and resources are available to us," he writes in *The Precipice*. "Our abiding image of space travel should not be the comfort and ease of an ocean liner, but the ingenuity, daring and perseverance of the Polynesian sailors who, a thousand years ago, sailed vast stretches of the Pacific to find all its scattered islands and complete the final stage of the settlement of the Earth. . . . If we could reach just one nearby star and establish a settlement, this entire galaxy would open up to us."[88] And while Ord doesn't think going to space would stop all existential threats, he does think it could make humanity safer than if we all stay on Earth. Once the colonization of our galaxy begins, he writes, "the process would be robust in the face of local accidents, failures or natural setbacks."[89]

For the longtermists, then, space isn't just the location of the limitless future; it's also part of the solution to the risk of human extinction. But on top of all that, the longtermists claim that we *need* to get to space—that there is "a moral case for space settlement," as MacAskill puts it—and that failing to do so would be a cosmically grave tragedy.[90]

* * *

Looking out into space forces us to confront a surprising fact about ourselves. Our eyes, mine and yours, are time machines. You don't see things as they are now—you see them as they were. It takes a fraction of a second for your brain to perform multiple miracles of signal processing at once, managing the impulses from the light-sensing cells in your retinas and forming them into a single image of the world around you, incorporating new information from your eyes dozens of times a second. You're always a little more than ten milliseconds behind events in front of your nose. Yet that's just the start—go well beyond your nose, and things get truly delayed, thanks to the limited speed of light. The silvery crescent Moon, hanging lovely and delicate in the darkening azure sky at dusk, is an image more than a full second in the past. The blazing Sun that rises the next morning is already eight minutes out of date by the time you see it. The rest of the night sky is even older news. Mars is usually about twelve light-minutes away; Jupiter, about forty. We see Sirius, the brightest star in the night sky, as it was over eight years ago. But when we look at Betelgeuse, the tenth-brightest star, we see it as it was over five hundred years ago, when dodo birds still roamed across a tropical island in the western Indian Ocean not yet given the name Mauritius. And the light from the Andromeda galaxy, the most distant object visible with the naked eye, is two-and-a-half million years old—older than humanity itself, when our ancestors were just learning to fashion stone tools on the plains of East Africa.

But the oldest light in the universe dwarfs even the age of the Earth. Find an old TV set with an actual antenna—maybe in the basement? The thrift store?—and turn it to a dead channel, filled with static. Turn the brightness down low, and watch the little blips of gray flash against the dark screen. About 1 percent of those flashes are caused by the cosmic microwave background, the earliest light there is, hitting the antenna.[91] That light has been traveling for nearly fourteen billion years. In that time, the expansion of space has carried the light's point of origin about forty-six billion light-years away, to the edge of the observable universe. The light formed only a few hundred thousand years after the Big Bang, when the entire universe was still so hot from that violent event that it was glowing a dull red, half the temperature of the surface of the Sun today. The packet of unspeakably ancient light that hit your TV antenna was originally in the near infrared, just outside of the range of human vision. But on its long, epic journey, the expansion of the universe itself stretched out that light, sapping it of energy and ramping it down from the jittery warm infrared to a long, lazy radio wave. Time changes everything—even the first light.

That same cosmic expansion is also going to leave us quite lonely. The observable universe is over ninety billion light-years across, containing some two trillion galaxies, but not for long. The expansion of the universe is accelerating, as it has for the last few billion years. In roughly a hundred billion years, it will carry away all but the closest galaxies—our Milky Way along with Andromeda, Triangulum, and a large array of dwarf galaxies, collectively known as the Local Group. The rest will be gone forever, lost to our sight. Around that time, the entire Local Group will merge into one large galaxy, "Milkdromeda," and that will be it—the only galaxy we can see, anywhere.[92]

Yet even after the rest of the cosmos has abandoned us, we will still be left with a bounty almost beyond measure—for a time. There are well over a trillion stars in our Local Group, most of which have planets around them, over a million million worlds covered in frozen air or rock or gas or even water. There are vast nebulae, fields of gas light-years wide collapsing to form new stars, with strange new worlds coalescing around them out of specks of dust and rocks the size of Manhattan; there is a black hole lurking at the center of our own galaxy, toying with the stars around it, playing with its food, forcing blue supergiants to whip around at a few percent of the speed of light before their final, inescapable fall down its gravitational gullet. A hundred billion years from now, locked in the lonely depths of Milkdromeda, the stars will still shine. But they will die too, sooner or later: the larger ones greedily burning through their store of atoms in millions of years until they die as supernovas, each one briefly outshining the rest of the galaxy combined; the smaller ones carefully conserving their fuel over the course of billions of centuries, then shining a thousand times longer than that as white dwarfs before they fade to black. Their light will die. Their planets will be engulfed in flame or freeze in the depths of space. Their remnants will decay or be consumed by black holes, and then those black holes, too, will die, evaporating into radiation, fading into the long night. Ultimately, all that will remain is this radiation, mixed with ghostly echoes of the original cosmic background, stretched ever thinner by the relentlessly accelerating expansion of space, eating away at what little energy remains in the echoes of the Big Bang.

This is the heat death of the universe—the final death of all differences in temperature, of all differences of any kind, as dictated by the most unbreakable law of physics there is: the second law of thermodynamics, which states that entropy must always increase in a closed system like our universe. Entropy is often explained as a measure of disorder, but on these cosmic timescales, entropy can be thought of as a measure of sameness—if there's a difference between two places in the universe, increasing entropy will smooth it out over time until they're identical, equally filled with equally random stuff at equal temperature. This race to sameness runs the universe even today. The entropy of the cosmos relentlessly increases, through myriad paths both bizarre and familiar. The Sun radiates its heat away, warming its surroundings as its low-entropy nuclear fuel runs down; an ice cube melts in a glass of water until they are one and the same. The Earth takes in low-entropy light and heat from the Sun—high temperature, coming in from one direction—and reradiates it out at a higher entropy—lower temperature, sent in all directions. Plants take that low-entropy sunshine and use it to break and forge chemical bonds to make sugar, a relatively low-entropy fuel source, but they increase entropy overall by radiating out heat in the process. We take that sugar and consume it for fuel, all the while radiating out high-entropy body heat in all directions, all day, every day. Time passes, the direction of the future itself indicated by the direction of increasing entropy: the low-entropy state of the universe in the past is what makes the past different from the future, giving time its arrow. The future comes, one day at a time, marching the universe toward heat death, with its promise of higher entropy in all things great and small. Disorder builds up in our cells; we age, we die. The Earth dies. The Sun dies. Eventually, everything dies.

This is not the only possible fate for the universe. But it's arguably the most pleasant of the options available. All of them involve entropy inexorably increasing; they only differ in the duration of the universe as we know it. Some involve a re-collapse of the universe, leading to a reverse Big Bang; others literally tear the universe apart while the stars are still shining (though these are both highly unlikely). Most alarming, there is the possibility of vacuum decay: a sudden change in the underlying state of space-time in one region of the universe due to a random quantum fluctuation, creating a bubble expanding at the speed of light that destroys literally everything in its path, forever. The end of all things, instantaneously, with no warning. None of these possibilities have been ruled out. But heat death against a background of endless, accelerating expansion is widely regarded as the most likely fate for the universe by modern cosmologists—with a possible coda of vacuum decay in the deep future, well after the heat death has imposed its relentless sameness across the universe.

How should we feel about this? In the face of the death of all things, what do we do? How should we live? I don't know. We know, abstractly, that we will die. The end of humanity is even more abstract. To fathom the end of the stars themselves seems beyond our grasp. The times involved are so large that we can't understand them directly, and even analogy fails. The last black hole in the observable universe will evaporate no more than 10^{100} years from now or thereabouts; I have spent much of my career trying to seek out a way to intuitively explain the size of numbers that large, and I don't think there's a good way to do it. Baffled, then, seems like an appropriate response to the dark fate of the cosmos. So does despair, anger, and hope—hope that physics is wrong, that I am wrong. And perhaps I am wrong. But this is what the best science of today tells us is in store for the universe. We don't get to choose that. All we can choose is how we respond to it.

Allow me, then, to suggest a response. It's not the right response—there is no right response to the end of all things—but it is, I think, a healthy one. It starts with a story, apocryphal and of uncertain provenance. Several decades ago, a great astronomer was giving a public lecture about the solar system to a packed auditorium. She described how the Sun and planets formed 4.5 billion years ago out of a collapsing cloud of gas and dust, how the Sun's gravity keeps the planets in their elliptical orbits; and finally, how in five billion years, the Sun would expand and become a red giant, swallowing the Earth. At this last piece of news, a shaky hand went up from a man in the third row. The astronomer nodded at him, and he stood up.

"Excuse me, but did you just say the Sun would swallow the Earth in five million years?" he asked.

"Not quite—that will happen in five *billion* years," the astronomer answered.

"Oh, *thank goodness*!" the man replied, visibly relieved.

It's true that the universe will end. But for now, we are alive, here on Earth. This planet is our home. Billions of years of evolution have ensured our adaptation to it, its hospitability for us. There is natural beauty on this planet beyond anything the human imagination could devise. There are whales the size of passenger jets, tardigrades half the width of an eyelash, and gnarled trees hidden in the mountains that are older than the pyramids at Giza. There are snow-covered volcanoes in the permanent winter of Antarctica; there are rifts a mile under the Pacific teeming with life; there are clear rivers wandering through subalpine meadows, with water-smoothed rocks beside them that stay warm all afternoon in summer. There are also eight billion other people on this planet who need our regard, our care, and our love, right now. Our actions and feelings—the help we give each other, the pleasure we take in the splendor of this world—will not lose their meaning or value with the passage of time. The impermanence of the universe does not make existence meaningless—it will always be true that we were here, even after all trace of us has been erased. We are here now, in a world filled with more than we could ever reasonably ask for. We can take joy in that, and find satisfaction and meaning in making this world just a little bit better for everyone and everything on it, regardless of the ultimate fate of the cosmos.

That's one possible response. Here's another one:

> As I write these words, suns are illuminating and heating empty rooms, unused energy is being flushed down black holes, and our great common endowment of negentropy [low-entropy energy] is being irreversibly degraded into entropy on a cosmic scale. These are resources that an advanced civilization could have used to create value-structures, such as sentient beings living worthwhile lives. The rate of this loss boggles the mind. One recent paper speculates, using loose theoretical considerations based on the rate of increase of entropy, that the loss of potential human lives in our own galactic supercluster is at least $\sim 10^{46}$ per century of delayed colonization. . . . Even with

> the most conservative estimate, assuming a biological implementation of all persons, the potential for one hundred trillion potential human beings is lost for every second of postponement of colonization of our supercluster. From a utilitarian perspective, this huge loss of potential human lives constitutes a correspondingly huge loss of potential value. . . . The effect on total value, then, seems greater for actions that accelerate technological development than for practically any other possible action. Advancing technology (or its enabling factors, such as economic productivity) even by such a tiny amount that it leads to colonization of the local supercluster just *one second* earlier than would otherwise have happened amounts to bringing about more than 10^{29} human lives (or 10^{14} human lives if we use the most conservative lower bound) that would not otherwise have existed. Few other philanthropic causes could hope to match that level of utilitarian payoff.[93]

That's Nick Bostrom, in his 2003 paper "Astronomical Waste: The Opportunity Cost of Delayed Technological Development." Here's Toby Ord, from *The Precipice*:

> At the ultimate physical scale, there are 20 billion galaxies that our descendants might be able to reach. Seven-eighths of these are more than halfway to the edge of the affectable universe—so distant that once we reached them no signal could ever be sent back. Spreading out into these distant galaxies would thus be a final diaspora, with each galactic group forming its own sovereign realm, soon causally isolated from the others. Such isolation need not imply loneliness—each group would contain hundreds of billions of stars—but it might mean freedom. They could be established as pieces of a common project,

> all set in motion with the same constitution; or as independent realms, each choosing its own path. . . . If we can venture out and animate the countless worlds above with life and love and thoughts . . . we could bring our cosmos to its full scale; make it worthy of our awe. And since it appears to be only us who can bring the universe to such full scale, we may have an immense instrumental value, which would leave us at the center of this picture of the cosmos. In this way, our potential, and the potential in the sheer scale of our universe, are interwoven.[94]

And here's Anders Sandberg and Stuart Armstrong, formerly a researcher at FHI, in a 2013 paper on how, exactly, all this could be accomplished:

> Travelling between galaxies—indeed even launching a colonisation project for the entire reachable universe—is a relatively simple task for a star-spanning civilisation, requiring modest amounts of energy and resources. We start by demonstrating that humanity itself could likely accomplish such a colonisation project in the foreseeable future, should we want to. Given certain technological assumptions, such as improved automation, the task of constructing Dyson spheres, designing replicating probes, and launching them at distant galaxies, become quite feasible. . . . On a cosmic scale, the cost, time and energy needed to commence a colonisation of the entire reachable universe are entirely trivial for an advanced human-like civilisation. Once universal colonisation is en route, a second wave to colonise the galaxy could be launched on similar time scales. . . . Nearly every star can be grabbed in two or three generations of replicators.[95]

We've seen this before. These are essentially the same self-replicating constructor bots that Kurzweil predicted, paving over literally all of nature across the entire universe. And just like Kurzweil, these leaders of longtermism think that the lack of alien constructor bots is an indication that we're probably alone in the universe. "A star-spanning civilisation would find the energy and resources required so low, they could do this project as an aside to their usual projects. . . . This result implies that the absence of aliens is more puzzling than it would be if we simply considered our own galaxy," Sandberg and Armstrong wrote in 2013.[96] In a 2018 paper, Sandberg, Ord, and Eric Drexler concluded that the most likely explanation for this absence is that we're alone in the universe, or nearly so. "We find a substantial probability that we are alone in our galaxy, and perhaps even in our observable universe," they wrote, claiming that any alien civilizations are "probably extremely far away, and quite possibly beyond the cosmological horizon and forever unreachable."[97] Bostrom agrees. "My guess is that our observable universe doesn't contain intelligence," he says. "So we don't need to worry about taking matter away from them." Bostrom's vision for this is somewhat different from the details of Armstrong and Sandberg's ideas, but for him, the desired upshot is the same as Kurzweil's dreams: nature destroyed in the name of transforming the universe into a giant computer. He envisions "a growing sphere, a bubble of technological infrastructure with Earth at the center. It's growing in all directions at the speed of light. . . . Most likely, everyone would live in virtual reality, or some abstract reality."[98]

Yet despite the claim that these technologies are the unavoidable future of all intelligent life—more of the rhetoric of inevitability—the science behind the longtermists' case for space is shaky at best. Sandberg and Armstrong's scheme for colonizing the universe requires creating a fleet of extremely powerful self-replicating space probes—but to build those, we'd need an awful lot of technology that doesn't exist yet and might never exist. Kurzweil proposed similar probes using Drexler-style nanotechnology, but those kinds of molecular constructors and disassemblers face steep obstacles. There could be ways for Sandberg and Armstrong's self-replicating probes to achieve their goals without going all the way down to the molecular scale, but that would mean some material would remain unprocessed and thus "wasted," by the longtermists' lights. Sandberg and Armstrong also seem to imply that their probes would allow for uploaded minds to reside in the computers crossing the cosmos; Bostrom says it more explicitly. Yet mind uploading is also an entirely speculative technology—not to mention that Bostrom's estimate for the computational power of the human brain is itself controversial, as there's no consensus value for it (and not all neuroscientists agree that the brain's activity can be fully represented by computation in the first place).

Regardless of the viability of these specific technologies, self-replicating interstellar probes would need to be entirely self-contained. They'd be operating much too far away from home to be able to rely on commands from Earth. Given the complexity of their mission and the unpredictability of their surroundings, this would probably require each probe to house a full-blown AGI, capable of making its own plans and decisions. In any event, that's what Bostrom and other longtermists have in mind: they want to see a universe bristling with happy, conscious intelligence, even if it's not human. But AGI is itself yet another nonexistent and unproven technology. We don't know what it would require, nor whether there would be unforeseen barriers to housing an AGI in a probe like this and sending it across intergalactic distances. (And even if there isn't, there might be dicey ethical questions: Would an AGI really want to go on this phenomenally dangerous one-way trip? If it's a fully conscious and intelligent being, its opinions would need to be considered.) The same goes for all the other nonexistent technology that these probes would need to carry, like their wildly speculative propulsion systems, which, even if they can be created, might not work well in space or in such tight quarters. (The self-replicating probes would need to be pretty small, as we'll see in a moment.)

Even if such probes are technically feasible, though, building them could be a very bad idea. Self-replicating probes of the sort Sandberg and Armstrong envision would be extraordinarily destructive, with the ability to tear apart and consume entire planets. They would make nuclear weapons look like firecrackers. The creation of such a machine would itself constitute a major international relations crisis and could precipitate an arms race where one false move destroys the entire Earth, and possibly the rest of the universe too. Moreover, the longtermists plan to put AGIs into these probes, creating a question of trust between whoever creates the AGIs that go in the probes (perhaps an AGI itself) and the rest of humanity.[99] The existence of new devices that could rapidly destroy the entire planet in the event of just one accident (or just one unstable person getting a hold of the technology) seems like the kind of thing that people interested in avoiding the extinction of humanity should be advocating against, rather than working to hasten.

There's also the problem of the materials and energy necessary for these self-replicating probes to work. Sandberg and Armstrong propose launching these probes at somewhere between 50 percent and 99 percent of the speed of light, depending on the details. These are truly outrageous speeds. The top speed of the Parker Solar Probe, the fastest artificial object in history to date, is less than 0.1 percent of the speed of light. Sandberg and Armstrong are rather sketchy on the details about what kind of launch system might be able to propel these probes at such high speeds. They mention a few speculative systems, none of which has ever been used to launch anything like this, and some of which have never been built at all. But that's arguably not the biggest problem when it comes to launching these probes, because no matter how they're launched, it would require an enormous amount of energy to get them up to such high speeds. Sandberg and Armstrong's scheme for obtaining that energy is to "take Mercury completely apart" and use it to blot out the Sun's light with a "Dyson swarm."[100] This is a version of a Dyson sphere, an enormous structure that encloses a star in order to capture all of its energy and put it to use. Dyson spheres are named after the physicist Freeman Dyson, who wrote a paper about the concept in 1960, working off of an idea from the early twentieth-century sci-fi writer Olaf Stapledon. Such spheres probably couldn't be solid structures, despite the name. Instead, physics suggests that Dyson spheres could be built out of a swarm of small independent structures around the host star. Sandberg and Armstrong suggest placing such a Dyson swarm at the same distance from the Sun as Mercury; a sphere of that size would have a surface area of over four thousand trillion square kilometers, more than eighty million times larger than that of the entire Earth. Spreading the material of Mercury out across that area, Sandberg and Armstrong calculate, would mean the swarm couldn't be thicker than about half a millimeter. Packing in the materials needed to harness and use the energy of the entire Sun in such a swarm would be a challenge, especially with the tremendously high radiation levels hitting it from the nuclear fireball contained within; keeping the swarm safe from solar flares and storms would be even harder. But after disassembling an entire planet to build the swarm, sending out intergalactic probes to colonize the universe would be comparatively straightforward: it would only require using a small fraction of Mercury's material to build the fleet of probes. Then the Dyson swarm would capture the entire energy output of the Sun for a few hours to launch the probes at nearly the speed of light using a hypothetical propulsion system.

Even assuming these probes could be built and launched, there are still more barriers they would face. The voyage to another galaxy would be long, over two million years at least—or ten thousand times longer than that, depending on the target galaxy—though it would be significantly shorter than that from the perspective of the probes themselves, thanks to the magic of relativity. Riding aboard a probe traveling at 99 percent the speed of light, it would "only" take about 360,000 years for them to arrive at Andromeda. But they would pay a price for this magic trick: from their perspective, the rest of the universe would be whizzing past them at 99 percent of the speed of light, which means any isolated subatomic particle in deep space would become dangerous high-energy radiation, and any speck of dust would become a small explosive. Since most of the probes wouldn't make it to their destination as a result of near-light-speed impacts, Sandberg and Armstrong suggest sending redundant groups of probes to each galaxy. But that doesn't solve the problem that single subatomic particles could cause in the form of high-energy radiation along the journey, or similar problems that could be caused by lower-energy radiation once the surviving probes arrive at their destination. Radiation wreaks havoc on all known computing systems; it can cause bits to flip, memory to be corrupted, commands to go unrecognized. This is not desirable behavior in any computer, much less one attached to machinery that could destroy an entire star system.

If they do survive the journey, though, the probes would still be faced with the challenge of doing exactly that: destroying an entire star system. Their immediate goal would be the construction of another Dyson swarm, to harness the energy of a star in the target galaxy and use it to power the creation of another generation of self-replicating intergalactic and interstellar probes, with the ultimate goal of surrounding every star in the accessible portion of the universe to prevent Bostrom's "astronomical waste."

Sandberg and Armstrong are quick to point out that this scenario isn't necessarily what they think will actually happen. "It is not our contention that a universal colonisation is likely to happen in this exact way, nor that the whole energy of a star will be diverted towards this single goal (amusing as it is to picture a future 'President of the Solar System' proclaiming: 'Everyone turn off their virtual reality sets for six hours, we're colonising the universe!')," they write. "Rather this reveals that if humanity were to survive for, say, a million years—still a cosmically insignificant interval—and we were to spread over a few star systems, and spend nearly all available energy for our own amusement, we could still power these launches, many times over, from small amounts of extra energy." The point of all this theorizing isn't that this specific scenario will happen. They're just making the argument that sending out such probes across the universe is feasible by providing a plan for doing so that they claim is not just plausible, but "conservative." This is a bold claim, since none of this technology has ever been tested at all. There's no viable plan for building it, and it's entirely possible that it simply wouldn't work. Sandberg and Armstrong repeatedly point out that there's no known barrier to it from fundamental physics—which is probably true—but that is far from saying that no barrier exists, or that it would ever be possible to build such a system. There might be materials required to build such a system that don't exist, or can't exist. Without a more detailed plan and more testing, there's no way to know if such technologies are possible. A mere sketch of a feasibility argument doesn't mean that any of these technologies, or anything like them, could actually be built.

Yet even if this sort of future is technologically possible, there's still the question of whether such a future is desirable in the first place. The longtermists are quite optimistic that such a future is possible, but they're even more certain that it's desirable—so much so that they believe its desirability to be a universal quality, one that would be shared by any (or at least many) intelligent species anywhere in the cosmos. Hence Bostrom and Sandberg's belief that alien civilizations don't exist anywhere around us for billions of light-years, if at all. If such aliens did exist, we'd see their self-replicating probes, or at least evidence of their Dyson spheres in the form of stars winking out across the cosmos. Yet this argument assumes that such technologies are possible at all—perhaps the lack of self-replicating probes and Dyson spheres is simply evidence that *they can't be built*—and also makes an enormous number of unwarranted assumptions about what aliens would want and how they would see the world, a real failure of imagination. Aliens might be *alien*. They might have completely different sensory structures from ours. Things that seem obvious to us might seem abstruse or even wrong to them, and vice versa. They might be unrecognizable; they might not be able to move; they might not be made of anything solid; they might eat diamonds and excrete lava. Their motivations will be shaped to some degree by their physical form—which we can make some informed guesses about—but their motivations will also be shaped by their culture, and there we have absolutely no idea what might be going on. Human cultures vary so widely, and that's just within our own species. There are other intelligent creatures on Earth—dolphins, elephants, crows, octopuses—and their languages and customs are totally, wildly different from our own despite our shared heritage. Even chimpanzees and bonobos, our nearest relatives, are alien to us; they want different things than we do, and communicating with them is difficult and unreliable. How can we hope to guess what a creature that evolved entirely independently, in a wholly different ecosystem on another world orbiting an alien star, might think is important? They might not even have concepts like "importance" or "rationality" or "culture" or "space." Even most humans don't want to build a Dyson sphere or send out self-replicating probes. Why should anyone think that all—or even any—aliens would want to do those things?

For that matter, why do the longtermists want to do those things? If there are aliens—and there are good arguments that at least some forms of alien life exist, even if they're just the equivalent of bacteria—then paving over the universe is a crime against all the other life scattered across the sky. "If there is life on Mars, I believe we should do nothing with Mars," Carl Sagan wrote in *Cosmos*. "Mars then belongs to the Martians, even if the Martians are only microbes. The existence of an independent biology on a nearby planet is a treasure beyond assessing, and the preservation of that life must, I think, supersede any other possible use of Mars."[101] What Sagan said should go for any planet with any life on it at all: it should be left undisturbed, no matter how simple that life is. And even if there's no life at all anywhere else in the universe (which would be impossible to prove anyhow), then sending out self-replicating probes would still be a cosmic injustice, precluding the possibility of new life and new civilizations ever arising while simultaneously robbing the cosmos of any semblance of natural beauty. To MacAskill's credit, he has suggested that galaxies that have "even a 1 percent chance of naturally developing advanced life [should] be left alone." But this is hardly an answer, since such a determination would be impossible to make. It's also unclear what would constitute "advanced" life or why the "advanced" status of that life should matter. MacAskill also suggests that all the other galaxies in the universe should be split among all humans, as though we have a right to the entire universe.[102] There's something almost circular about the logic at work: we don't see self-replicating probes, so there must not be aliens, therefore we are morally justified in sending out self-replicating probes. Just on the face of it, it seems absurd. Destroying entire galaxies' worth of planets to rob them of their starlight sounds like an intergalactic James Bond supervillain scheme. Why do Oxford ethicists want this? Their stated motivation is to create as many people or sentient beings as possible before the heat death of the universe. But why? What's the point of this cruel dream? Do they think they're going to get a high score on the leaderboard at the end of the universe?

Apparently so. Many longtermists (including MacAskill) tend to favor a particular idea from utilitarianism called the "total view," which states that the best thing to do is to maximize the utility (whatever that is, but let's call it happiness for now) of the total population of humans or sentient creatures. So, for example, given the choice between a world with one extremely happy person in it and two people who are each only 75 percent as happy as that first person, someone holding the total view would choose the second option, because the total amount of happiness is 50 percent larger in the second world. This could continue: given the choice between a world with eight billion reasonably happy people and eighty trillion people who are each only one-thousandth as happy on average, the total view demands we pick the latter, because there's ten times as much total happiness in that world. Taking this to its logical conclusion, the total view implies that a universe filled to the brim with miserable people who are (on average) just barely happy enough to make their lives worth living is better than a world with a smaller number of genuinely happy people. This is known as the Repugnant Conclusion, a term coined by the influential twentieth-century philosopher Derek Parfit, who rejected the idea. He wasn't alone. "Many [ethicists] regard the Repugnant Conclusion as a refutation of the total view," writes the MIT philosopher Kieran Setiya. "Imagine the worst life one could live without wishing one had never been born. Now imagine the kind of life you dream of living. For those who embrace the Repugnant Conclusion, a future in which trillions of us colonize planets so as to live the first sort of life is better than a future in which we survive on Earth in modest numbers, achieving the second."[103]

Parfit, whom Setiya calls "iconic," was Toby Ord's PhD adviser, and he and many other longtermists hold Parfit's work in high regard.[104] A large chunk of MacAskill's book *What We Owe the Future* is about Parfit's ideas. MacAskill calls Parfit "one of the most creative and influential moral philosophers of the last century" and attributes many of the crucial insights that led to longtermism to Parfit, especially when it comes to thinking about existential risk.[105] But MacAskill, Ord, and many others in the EA movement explicitly disagree with Parfit about at least one major issue: they are comfortable with the Repugnant Conclusion. "In what was an unusual move in philosophy, a public statement was recently published, cosigned by twenty-nine philosophers, stating that the fact that a theory of population ethics entails the Repugnant Conclusion shouldn't be a decisive reason to reject that theory," MacAskill wrote in *What We Owe the Future*. "I was one of the cosignatories."[106] Ord, Greaves, and two employees of Open Philanthropy were also among those who signed it. "The Repugnant Conclusion depends crucially on intuitions about cases with very large numbers of people. The size of such very large numbers is hard to grasp on an intuitive level," they wrote. "We are also bad at compounding small numbers. We may therefore fail to see how lots of lives with a small but positive value could add up to something very valuable."[107]

The Repugnant Conclusion does seem repugnant on its face. But the longtermists' willingness to entertain it anyhow—MacAskill writes it's a view he's inclined toward—forces us to ask why it's repugnant, where the discomfort with the conclusion comes from.[108] Trillions of people barely happy enough to keep on living just doesn't seem like a good way for the world to be, not when given the alternative of billions of genuinely happy and fulfilled people living their best lives. There's something missing here, something that is lacking in the total view and its perspective of ethics, which is reflected in the Repugnant Conclusion. A twist on the Repugnant Conclusion's thought experiment suggests part of the problem. Imagine a town, Rivertown, with ten thousand reasonably happy people, and another town, Lakeville, with ten thousand residents of its own, each one 80 percent as happy as the people in Rivertown. Now add to Lakeville a hundred more people, each of whom are incredibly, deliriously happy, with no cares and no work, all of their desires and whims tended to by the other ten thousand people in their town. Each of these elite one hundred are as happy as twenty people from Rivertown combined. Which town would you rather live in? The total view says it's a coin toss; both towns are equally good. Now consider a third town, Omelas, also home to ten thousand people, all of whom are incredibly happy.[109] Let's even say that they're each as happy as one of the elite one hundred from Lakeville. But there's also one more person who lives in Omelas, a six-year-old, and that child is sickeningly miserable: kept bound in a room, living in squalor, deprived of all human contact. They've never even seen the sky. They're well below the point where their life is worth living; their level of misery is so extreme that their negative utility cancels out half of the contributions from the rest of the residents of Omelas combined. The child doesn't want to live anymore, but they're kept alive anyhow, because the rest of Omelas needs them. They must be kept in their horrific conditions in order for the rest of Omelas to function. The total view says Omelas is a far nicer place to live than Rivertown or Lakeville. In fact, it's ten times better than either of the other towns.

Ord is quite explicit about this. He contends that a sufficiently large amount of joy for a large enough number of people can outweigh the suffering of one—or many. "The goodness of *stopping* a million people suffering in agony is only finitely many times as good as a happy year of life," he writes. "There must therefore be an amount of happiness so valuable that it is more valuable than avoiding a million people being in extreme agony."[110]

There is an air of the absurd about these mathematical calculations. Yes, the insensitivity of the total view to inequality, like that found in Lakeville and Omelas, is a serious flaw. But even more striking than the troubling moral arithmetic at work in the previous examples is the sense that no such calculations could actually capture what it means to be in such circumstances. How can we quantify the misery of the child in Omelas? How can we add up the happiness of all the residents of Rivertown and arrive at a single number? What units would that number be expressed in? Utilitarianism of this sort relies on there being a single thing called "utility" that can be defined and measured (in theory if not in practice) for every person. The total view further requires that utility measurements can be collected and added up across entire populations of humans. But it's unclear what kind of thing utility could be; gallons of ink have been spilled for centuries by philosophers trying to define it. Utility isn't simple pleasure. Enjoying the satisfaction of achieving a difficult but worthy goal (successfully climbing up a steep hill, or creating a work of art that moves someone else) is quite different from pure euphoria, and both add to the joy and rich texture of a good life. There's the quiet pride of seeing something you made used the way you'd hoped it would be; there's the way you feel seen when a friend takes the time to give you good advice at the right time; there's the uncut animal rush of plunging down a roller coaster. Any proposed measure for utility has to account for all these and more, and convert them into a single number. It also has to find a way of balancing happiness with misery, detracting from the total every time some flavor of bad or painful thing occurs to someone—everything from a stubbed toe to genocide. If someone told you that they had found a way to do that, one that worked for literally every human that had ever lived or could ever live until the end of the universe, would you believe them?

Yet there's an even deeper problem here, one that would arguably remain even if someone really could define utility in a satisfying way. "One of the objections that John Rawls [the twentieth-century political philosopher and ethicist] famously made about utilitarianism is its conception of persons, which it basically reduces to containers of value, receptacles of value," says Émile Torres, a philosopher at Case Western Reserve University specializing in the ethics of human extinction risk. "And so we're a means to an end. We're just the receptacles. Increase the number of receptacles that have net positive value and you increase the total amount of value in the universe, and that's good. And I just don't think of people that way. . . . In my view, people are just ends, and treating them as mere means is deeply problematic."[111] People are best seen as people, not as reservoirs for holding an abstract quantity of dubious ontic status.

The whole project of utilitarianism smacks of a kind of ethical Taylorism: confusing metrics with the reality they imperfectly capture, and then trying to optimize things by focusing solely on increasing the metrics to the exclusion of all else. Ethics isn't economics. And as Torres alludes to, utilitarianism isn't the only game in town when it comes to ethics. Utilitarianism is just one type (or really a subfamily) of a wider group of ethical views known as consequentialism. Consequentialists hold that the ethical value of actions resides solely in their consequences (hence the name). But consequences can take many forms. For example, rights consequentialists say that people (and other entities) have rights that shouldn't be violated, and the morality of an action depends on whether and how anyone's rights are infringed upon. Moreover, consequentialism itself is just one of several major options. Another is deontological ethics, championed most famously by Immanuel Kant. Deontological ethics says that an action is ethical if it's in accordance with universal ethical laws. Those laws present a problem for deontological ethics that is somewhat analogous to the problem of defining utility for utilitarians: What are those laws, how could we ever discover them, and how could they hold in all possible situations?

These questions arise because ethics is very hard, which is why it's been debated by extremely smart philosophers for thousands of years. It isn't easy to know how to be a good person, nor to figure out the right thing to do in a given situation. There's a zoo of other schools of thought on ethics, a huge variety of positions that have been worked out by brilliant philosophers and thinkers. For any given moral dilemma, there are dozens of different perspectives that could be taken and plausibly defended, with centuries if not millennia of literature to back up each one. It's hard to know how to be good, even if you want to be.

Any decent philosopher knows there's a chance they're wrong, especially when wandering into what Setiya calls "the deep, dark waters of population ethics"—the branch of moral philosophy dealing with questions about how many people there should be.[112] The total view is just one view on population ethics, and among philosophers it's not a particularly popular one. Even if it were, "you can't outvote an objection [like the Repugnant Conclusion]," writes Setiya. "There are profound divisions here, not just about the content of our moral obligations but about the nature of morality itself."[113] And MacAskill knows he could be wrong. "I think the balance of arguments favours the total view, but given how difficult the subject matter is, I'm not at all certain of this. Indeed, I don't think that there's any view in population ethics that anyone should be extremely confident in," he writes.[114] MacAskill claims he's not even sure about utilitarianism as a whole. "I'm not a utilitarian because—though it's the view I'm most inclined to argue for in seminar rooms because I think it's most underappreciated by the academy—I think we should have some degree of belief in a variety of moral views and take a compromise between them."[115] But, he writes, "we need to know how to act despite our uncertainty." MacAskill suggests handling this in a rather utilitarian style: assigning probabilities to different ethical views and then taking the weighted average among them to get the expected value.[116] The problem with this approach is that it puts a thumb on the scale for MacAskill's favored conclusions, as Setiya points out. "There is a threat that longtermist thinking will dominate expected value calculations in the same way as tiny risks of human extinction. If there is even a 1 percent chance of longtermism being true, and it tells us that reducing existential risks is many orders of magnitude more important than saving lives now, these numbers may swamp the prescriptions of more modest moral visions," he writes.

"The theoretical problem is that we ought to be uncertain about this way of handling moral uncertainty. . . . There's no way to insure ourselves against moral error—to guarantee that, while we may have made mistakes, at least we acted as we should, given what we believed. For we may be wrong about that, too."[117]

Luckily, MacAskill has hopes of solving this problem as well. In a short follow-up to *What We Owe the Future*, MacAskill imagines what a good future would look like for longtermists. After eliminating war, but before embarking on colonizing the billions of galaxies that have somehow been deemed unlikely to develop advanced life, his imagined future society briefly pauses to solve all of ethics. "Progress was faster than expected," MacAskill writes. "It turned out that moral philosophy was not *intrinsically* hard; it's just that human brains are ill-suited to tackle it. For specially trained AIs, it was child's play."[118] MacAskill doesn't explain why this is remotely plausible. "Moral judgment is one thing; machine learning is another," writes Setiya. "Whatever is wrong with utilitarians who advocate the murder of a million for a 0.0001 percent reduction in the risk of human extinction, it isn't a lack of computational power."[119]

MacAskill, to his credit, does seem to understand that, even in his imagined utopia, there would be some people who didn't believe his specially programmed, super-ethical AIs. "No moral arguments were convincing to everyone, however, and the world did not converge on a single moral view," he writes. "But this did not mean that conflict between competing moral views was inevitable. The universe was big—twenty billion affectable galaxies—and everyone's wildest imaginations could be realised a thousand times over. We could have it all."[120]

MacAskill should know this is false. The universe isn't even big enough for the wildest imaginings of the people who work in his building, much less all of humanity. Sandberg, Armstrong, and Bostrom's program of self-replicating probes and Dyson spheres to replace the natural world with an artificial one stems directly from their ethical views. Those views could be wrong. Yet if they manage to achieve their goals—if the myriad unlikely technologies required for the future they seek all turn out to be possible—their ethical views would be pushed out into the cosmos at large, in a way that would make them unlikely to ever be replaced or supplanted by other ideas. The longtermists seem quite happy to gamble the fate of the entire universe on their view of ethics being correct. If they're right about the technology but wrong about the ethics, they will have destroyed the universe for nothing at all.

This points to a final problem: longtermism's promise of enormous utility in the future can be used to justify horrors. "[In] the history of utopian movements that became violent . . . [many] have two things at their core," says Torres. "One is the obvious, which is this utopian vision of astronomical or infinite amounts of value in the future. And the second is a broadly utilitarian mode of moral reasoning. So ends justify the means. What are the ends? The ends are *freaking utopia*. What means are off the table, if I have to kill a million people to get to utopia? . . . We're talking about utopia, the stakes are so huge, that the ends justify the means."[121] History—especially the history of the twentieth century—is littered with famous examples of utopian movements turned violent. "To make mankind just and happy and creative and harmonious for ever—what could be too high a price to pay for that?" wrote Isaiah Berlin in 1990. "To make such an omelette, there is surely no limit to the number of eggs that should be broken—that was the faith of Lenin, of Trotsky, of Mao."[122]

A version of this utopian, extremist thinking is already at work among the rationalists: remember Yudkowsky advocating that world leaders risk nuclear war in pursuit of his goal. The leaders of the effective altruist movement—MacAskill, Ord, and others—have said repeatedly that they do not believe the ends justify the means, that longtermist reasoning shouldn't be used to conclude that it's OK to violate other people's rights or do harm. But an obsessively quantitative focus on ultimate outcomes is implicit in the evaluation of risks that Ord and MacAskill have both pushed, which ranks AI alignment above addressing global warming. Global warming will disproportionately affect poor people of color. But to the longtermists, any problem that promises an outcome short of full extinction isn't as important as something that could wipe out the entire species, even if the former is real and present and the latter is purely hypothetical.

If we disregard calamities that fall short of full extinction, wide swaths of human culture and diversity will be lost forever, eternally absent from the longtermists' glorious future in space. Who gets to decide what makes the cut? The entire EA movement, and especially longtermism, has a very specific idea of what matters. By looking only at what they consider most important, they ignore the other needs and problems in the world, all while claiming they're saving the species. Writing about MacAskill's first book *Doing Good Better*, in 2015, the Oxford philosopher Amia Srinivasan cut to the core of this problem with EA, and foresaw in part where it would go over the next decade. "MacAskill does not address the deep sources of global misery—international trade and finance, debt, nationalism, imperialism, racial and gender-based subordination, war, environmental degradation, corruption, exploitation of labour—or the forces that ensure its reproduction," she wrote. "Effective altruism doesn't try to understand how power works, except to better align itself with it. In this sense it leaves everything just as it is. This is no doubt comforting to those who enjoy the status quo—and may in part account for the movement's success."[123]

Indeed, it's hard not to see a conflict of interest arising from the lavish tech funding for EA. "Who doesn't want to believe that their work is of overwhelming humanitarian significance?" writes Srinivasan.[124]

Rather than simply doing good in the world, Silicon Valley may be using EA to buy an image of moral rectitude. And EA, in turn, is compromised by its reliance on philanthropy from the winners of the tech industry lottery. "Billionaire philanthropy is an exercise of power by wealthy people to direct their private assets toward some public influence," political scientist Rob Reich tells me. "When people exercise power in a democracy, it deserves our scrutiny, not our gratitude."[125] Reich is very familiar with EA. He knows many effective altruists, he's taught them at Stanford, and he served on the board of the EA nonprofit GiveWell from 2013 to 2019, during which time that organization helped to launch Open Philanthropy. "I consider myself a kind of friendly critic of much, although not all, effective altruism," he says. And they're right about some things. "It's true that much philanthropy or charity in the United States and in other countries does not go to address disadvantage, or repair [or] ameliorate poverty," he says. But, he continues, "to put it really, really bluntly, effective altruism, when it comes to longtermism, ditches the effectiveness and is only about the altruism for future generations." Echoing Srinivasan, Reich thinks that EA has a serious problem with its understanding of the world, especially regarding political and social power. "There's a kind of cultishness and a sense of zealotry on the inside," he says. "I don't begrudge anyone who takes moral commitments with utter seriousness, but the absence of regard for the wider world, a disinterest in understanding how power works more generally, and a baseline arrogance or hubris—the kindest thing one could say about it is, it's tiresome."[126]

Out of this hubris comes a blinkered view of the world. "MacAskill is evidently comfortable with ways of talking that are familiar from the exponents of global capitalism: the will to quantify, the essential comparability of all goods and all evils, the obsession with productivity and efficiency, the conviction that there is a happy convergence between self-interest and morality, the seeming confidence that there is no crisis whose solution is beyond the ingenuity of man," writes Srinivasan. "There is a seemingly unanswerable logic, at once natural and magical, simple and totalising, to both global capitalism and effective altruism."[127] At the core of that logic, for both capitalism and effective altruism, is the need for quantification. Any human activity that can be quantified is grist for the optimizing machinery of this worldview, and anything that can't be quantified is dismissed as unimportant. This is how the longtermists, ultimately, are forced to see people: as numbers. And those numbers, in turn, need to be maximized and optimized, so they can be plugged into the grand longtermist plan to squeeze as much utility as possible out of the universe before its inevitable end.

* * *

There's a frantic air about all this, the urgent need to go, go, go, to grab as much low-entropy matter and energy as possible before the end of the show trillions of years from now. And it's hard to avoid the suspicion that the drive behind that frenetic pace is fear: fear of an end, fear of death. That's nothing new either. It's natural to be afraid of death; immortality is one of humanity's oldest fantasies. And the engraved medallion around Sandberg's neck projects a matter-of-fact confidence in a hopeless dream of a very specific kind of life after death. As I stared at it, I kept thinking about something David Gerard, an author and critic of the rationalists and effective altruists, had told me a couple of days before. A friend of his had been swept up into the rationalist community some years earlier, and that friend had been interested in cryonics as a way of escaping death. "He was the sort of guy who was always looking for a cult. And he found one which he slotted into and absolutely was the thing for him," Gerard recalled. "He finally did sign up for cryonics and put the money in, and I congratulated him. Because at that point, it's a religious thing. And if someone has a religious achievement, and they're very happy about it, then I am extremely happy for them."[128]

After speaking with Sandberg, I left Trajan House, wandering back toward the center of Oxford. I crossed the street and went around a hedge and through a gate in a low fence. I made my way through a field overgrown with grass swaying lightly in the breeze, smelling of late spring, a high insect buzz cutting across the afternoon sunshine. Dotted through the grass were rounded gravestones, knee height, splayed at odd angles and covered in patches of lichen. The names were nearly worn smooth; the dates were a century gone or more. Over the fence, on the other side of Mill Street, Trajan House sits, gleaming with glass and steel, its occupants dreaming of myriad ageless silicon lives in a universe deprived of nature. But across the street from the immortalists, death waits for them in the tall grass.

5

DUMPSTER FIRE SPACE UTOPIA

Marc Andreessen wants the respect he's due. "Technology is the glory of human ambition and achievement, the spearhead of progress, and the realization of our potential," Andreessen wrote in 2023. "For hundreds of years, we properly glorified this—until recently."[1] Andreessen has been one of the central figures in the tech industry for more than thirty years. In early 1993, Andreessen, an undergraduate at the University of Illinois, Urbana-Champaign (UIUC), suggested a new feature in a software project out of CERN that was only about two years old at the time, known then as the "WorldWideWeb."[2] Specifically, he suggested an addition to hypertext markup language—aka HTML, the language websites are written in to this day—to allow the in-line display of images on WorldWideWeb pages. Andreessen and Eric Bina, a programmer at the National Center for Supercomputing Applications at UIUC, incorporated this feature into a new Web browser they'd been working on, called Mosaic.[3]

Andreessen and Bina's innovation launched the modern Web. Up until that point, a few people thought that the Web might be better than contemporary, now-archaic internet protocols like Gopher, wrote the engineer Robert Metcalfe, but upon the launch of Mosaic, "several million then suddenly noticed that the Web might be better than sex."[4] The next year, Andreessen and his business partners started a new company, later dubbed Netscape. Netscape Navigator, the new Web browser created at that company, became explosively popular. Netscape dominated the early Web through most of the 1990s, its users surfing between sites over the pings and whistles of dial-up in their homes; the company was an icon of the decade's dot-com boom. Ultimately, Netscape fell to Microsoft's Internet Explorer around the end of the nineties in the first round of the "browser wars," only to rise again in open-source form in the early 2000s as Mozilla Firefox.

By then, Andreessen had made his millions and was ready to turn them into billions. In 2009, Andreessen and Ben Horowitz, one of his former employees at Netscape, launched a venture capital firm, Andreessen Horowitz, commonly abbreviated as a16z. Their firm rapidly became one of the most well-known and successful tech VC firms, a position they hold to this day. They were early investors in Airbnb, GitHub, Instacart, Instagram, Oculus, Pinterest, Skype, Slack, and many more. Andreessen himself sits on the boards of several other tech companies, including Meta, the parent company of Facebook, and Coinbase, the largest cryptocurrency exchange in the United States. There is almost no person on the planet who has not been touched or affected in some way by Andreessen's work, or that of the firm he still helps to lead. He is, by any definition, Silicon Valley royalty, among the most connected and influential people in the entire industry—indeed, in the entire world—with an empire of resources and people at his back. Anything he says or does is instantly news. But he tends to shun traditional news media. Instead, he uses the a16z website as a mouthpiece, occasionally dropping interviews or essays there. And in

October 2023, he posted something else: a "Techno-Optimist Manifesto," pushing the idea that technology could cure essentially all the world's ills.

Andreessen's manifesto attempts to set itself up in opposition to Yudkowsky's rationalists, effective altruists, and others deeply worried about existential risk from superintelligent AGI. He claims that slowing down progress on AI is "a form of murder," because he thinks an advanced AGI would inevitably save lives. His manifesto even contains a list of enemies; the first entry on that list is "existential risk." But for all that Andreessen wants to draw a distinction between his own views and those of the rationalists and effective altruists, he can't quite get out of their frame of mind. He accepts nearly all the same premises they do; he just uses them to reach different conclusions. (Even his claim that slowing down AI research is a form of murder looks like a small twist on Roko's basilisk.) Andreessen claims that an intelligence explosion is coming, and he cites Kurzweil and his law of accelerating returns by name. He even seems to agree with the longtermists that more people in existence is always good—and he certainly agrees with them about the best fate for the universe. "Our planet is dramatically underpopulated,"* he claims, contending that Earth could "easily" support more than fifty billion of us, and that humanity's numbers will surge far beyond that as we take to the stars. Like Kurzweil, Andreessen doesn't believe we have much of a choice about this. The only alternative he sees is annihilation. We must expand forever or fade away. "Not growing is stagnation, which leads to zero-sum thinking, internal fighting, degradation, collapse, and ultimately death."[5]

---

* Andreessen also claims in his manifesto that the number of people in the world might already be going down, which is simply false. There's no serious debate on this, and the statistics on global population growth are very easy to find. Yet despite the seemingly willful ignorance behind that statement, Andreessen is not alone among tech billionaires in this specific false belief. Elon Musk has made similar statements, as have the "pro-natalist effective altruists" Malcolm and Simone Collins. This panic over a nonexistent crisis is intimately linked with both eugenics and the growth-at-all-costs mindset.

To fuel this eternal expansion, Andreessen wants humans to use more energy than we currently do—much more. "We should raise everyone to the energy consumption level we have, then increase our energy 1,000x, then raise everyone else's energy 1,000x as well."* Andreessen isn't concerned about what these increases in consumption (and population) might do to the environment. He thinks that technology is an environmental panacea; "sustainability" is on his enemies list. "Risk management," "trust and safety," "stakeholder capitalism," and "tech ethics" are on the list too, along with "social responsibility." Andreessen opposes any restrictions whatsoever on the power of markets, corporations, and capital. He claims that developing new technology under no-holds-barred capitalism is "*inherently philanthropic*" and that the most important thing is to develop more technology as quickly as possible. Andreessen—like Kurzweil, Yudkowsky, and the rest—doesn't spend any time thinking about how to direct such innovation, nor does he talk about how to promote and encourage technological development in any specific way, though one gets the sense that he's not interested in the kinds of government investments that built Silicon Valley into a corporate powerhouse in the first place. Instead, he's just interested in going faster. He wants Kurzweil's exponential curves to fuel the growth of technology and capitalism until the end of time. Andreessen even talks about an eternal future of economic statistics, claiming that "since human wants and needs are infinite, economic demand is infinite, and job growth can continue forever."[6]

This vision of perpetual job creation past the heat death of the universe looks like the effective altruists' intergalactic empire as seen

---

* At no point in his manifesto does Andreessen clarify who is part of his rhetorical "we," which makes this passage somewhat ambiguous. Does he want to raise all of humanity's energy usage up to that of the average American, or that of the average employee of a16z, or his own? It's hard to avoid the suspicion that this ambiguity reflects a fundamental vagueness and lack of clarity in Andreessen's thinking.

through the fun-house mirror of start-up culture. And that's exactly what it is. Andreessen's manifesto ends with a list of "Patron Saints of Techno-Optimism"; unsurprisingly, Kurzweil is on that list alongside quite a few libertarians. (There are also some more surprising and nonsensical entries, like Bertrand Russell and Andy Warhol.) But two of the most telling entries on the list are pseudonymous Twitter accounts: @bayeslord and @BasedBeffJezos (aka Guillaume Verdon, an AI start-up founder), two of the originators of the nascent "effective accelerationism" movement.[7] "Technology and market forces (technocapital) are accelerating in their power and abilities. This force cannot be stopped," they wrote in their "inaugural" list of principles, posted on Substack in 2022. "Effective Accelerationism, e/acc, is a set of ideas and practices that seek to maximize the probability of the technocapital singularity, and subsequently, the ability for emergent consciousness to flourish."[8] They even claim that statistical thermodynamics demands that "humanity solves problems through technological advancement and growth," simply as "a consequence of our physical reality."[9] (This is, frankly, nonsense. The opposite is true. Thermodynamics and statistical mechanics are the ultimate source of nature's implacable demand that growth always ends.)* Effective accelerationists have set themselves in opposition to AI "doomers" like the rationalists and effective altruists—describing themselves as "EA, but based"—while acknowledging their shared desire for an ever-growing technological utopia in space.[10] "Stop fighting the thermodynamic will of the universe.[11] You cannot stop the acceleration. You might as well embrace it. A C C E L E R A T E."[12]

Effective accelerationism has proven relatively popular among start-up founders, venture capitalists, and certain segments of the

---

* Any decent textbook on the subject makes this clear; a particularly good one is *Statistical Mechanics: Entropy, Order Parameters, and Complexity*, by James Sethna. "Entropy provides fundamental limits," writes Sethna, including limits on life, computation, civilization, and ultimately the entire universe.

extremely online. Some have taken to putting "e/acc" into their social media profiles—Andreessen has done this, putting him in the company of the infamous "pharma bro" Martin Shkreli and, for some reason, nineties rapper MC Hammer. Many of the ideas espoused by Andreessen and the effective accelerationists are taken quite seriously among other tech billionaires too. Sam Altman believes that AGI is coming, and, like Andreessen and Kurzweil, he believes this will lead to "Moore's Law for everything."[13] That's the mechanism by which OpenAI will, supposedly, obtain all the wealth in the world: a privately owned Singularity, with all the wealth accumulating to the one company that created the AGI that kicked off the intelligence explosion. Altman thinks this will happen soon. "If you believe what I believe about the timeline to AGI and the effect it will have on the world, it is hard to spend a lot of mental cycles thinking about anything else," he said in 2019. "So I have not thought deeply about what it would take to solve, really, any other problem in the last few years."[14]

Altman said this in an interview with the economist Tyler Cowen, and the context is especially revealing. Cowen hadn't asked him about AI. He was, instead, asking him about housing policy in and around San Francisco. Altman prefaced his reply with his statement about AGI as a way of explaining why he hadn't thought much about housing policy, but that answer also implies that anything other than AGI is ultimately just a footnote. This is a key part of the appeal of the ideology of technological salvation, one that's especially important to Silicon Valley billionaires and tech executives: boiling all the problems in the world down to one question of computer technology. Altman even claims that saving democracy itself requires the growth that tech start-ups will purportedly enable. "Democracy only works in a growing economy. Without a return to economic growth, the democratic experiment will fail."[15] And in 2023, Altman and Ilya Sutskever said that AGI would solve global warming. "I don't want to say this because climate change is so serious and so hard of a problem, but I think once we have a really powerful superintelligence, addressing climate change will not be particularly difficult for a system like that," Altman said. "I think this illustrates how big we should dream. You know, if you think about a system where you can say, 'Tell me how to make a lot of clean energy cheaply,' 'Tell me how to efficiently capture carbon,' and then 'Tell me how to build a factory to do this at planetary scale'—if you can do that, you can do a lot of other things too."[16] Altman is so confident in this "plan"—solving global warming by asking a nonexistent and ill-defined AGI for *three wishes*—that he's willing to gamble our climate and our future on it. "We do need way more energy in the world," he says. "AI is going to need a lot of energy. . . . I still expect, unfortunately, that the world is on a path where we're going to have to do something dramatic with climate, like geoengineering as a Band-Aid, as a stopgap."[17] Bill Gates has also long been a proponent of geoengineering as a solution to climate change, despite its massive risks.[18] (It's hardly surprising to see Altman and Gates singing the same tune, since their fortunes are aligned through OpenAI's contract with Microsoft.) It's a convenient idea: permission to use as much energy as your companies need, regardless of the ecological impact, in service of a higher cause that's always just over the horizon.

Meanwhile, other tech billionaires are working to build an intellectual movement to underpin this notion of growth at all costs. "One of the most important facts in the world and the history of civilization to date is that the rate of progress has not been constant," said tech billionaire Patrick Collison in 2019. "How does progress happen, how do we discover useful knowledge, how is that diffused, and how can we do it better? . . . There is a moral imperative to this kind of progress, and we shouldn't lose sight of that fact."[19] Collison founded the payment processing company Stripe with his brother John in 2010; he is among the youngest billionaires in the world, with a net worth estimated at

$7.2 billion. In 2019, Collison and Cowen wrote in an essay about the need for a new "science of progress," which "would study the successful people, organizations, institutions, policies, and cultures that have arisen to date, and it would attempt to concoct policies and prescriptions that would help improve our ability to generate useful progress in the future."[20] To that end, the Collisons have funded several different projects, including Roots of Progress, a nonprofit devoted to "establishing a new philosophy of progress" based on "a clearer understanding of the nature of progress, its causes, its value and importance, how we can manage its costs and risks, and ultimately how we can accelerate progress while ensuring that it is beneficial to humanity."[21] The Collisons were also involved in the creation of a new division at Stripe, Stripe Press, a media and publishing outfit pushing similar ideas about the future of technology, progress, and growth. Stripe also publishes *Works in Progress*, a magazine about "new and underrated ideas to improve the world."[22]

One of those "underrated" ideas is using as much energy as possible. "I have a confession: I'm an ergophile. I love energy-intensive processes and increasing the amount of power (in the physics sense) each of us can access," wrote Benjamin Reinhardt in *Works in Progress* in an October 2022 article titled "Making Energy Too Cheap to Meter." "In the extreme, energy is the *only* scarce resource. . . . To cap our energy ambitions is to commit to permanent scarcity."[23] Reinhardt—who describes his job as "working on how to enable more amazing sci-fi things to become reality"—used to believe in sustainability.[24] "As a California child of the 1990s, I was raised on *Captain Planet*, *Fern Gully*, and parking-meter-like machines that let you deposit a quarter to save a tree in the Amazon," he wrote. "It was clear that energy-intensive processes meant rapacious industry stripping pristine environments, birds dying in black gunk, decay, radioactive waste, smokestacks, war over burning oil fields, and an overheating planet."[25]

But then, he said, "what changed was me just understanding how the world works."[26] Using more energy, according to Reinhardt, means we could raise everyone's standard of living. Anticipating Andreessen's manifesto—which was published about a year later—Reinhardt wrote in *Works in Progress* that "sustainability . . . means a fixed pie, and the conflicts that inevitably erupt from it. . . . A renewed trend of exponential energy can both solve problems of the past and enable so many possibilities for human flourishing." Using more energy, he claimed, "is not only compatible with concerns about the environment and humanity, but critical for a flourishing future where we aren't fighting over a fixed pie. . . . Instead of [the outsize energy usage of Americans] being a source of shame, let's use them as a North Star. An ergophilic world is the only way that everybody can have the quality of life that we few enjoy today." Like Andreessen, Reinhardt isn't worried about environmental effects, because, he claims, the solution to such problems "is, counterintuitively, *more energy*." Using more energy creates more waste, but Reinhardt writes that "more energy can solve waste problems as well: At high enough temperatures, everything breaks down, so trash and harmful waste problems could disappear."[27]

This sounds like Reinhardt is proposing that all waste be burned, and when I ask for clarification, he tells me that's exactly what he has in mind. "This is not a super weird, futuristic thing. There are high efficiency incinerators. I lived in Singapore for a little over a year, and that's what they do with their waste," he says. "They just have these super high-efficiency incinerators and just throw everything in there. I don't know exactly how it works, but my understanding is that it is very, very clean. . . . I assume that it produces some amount of volatile gases and charcoal, and probably some metal slurry that runs off."[28] If Reinhardt had looked into it more, he might have discovered that the trash incinerators in Singapore are not quite as clean as he'd hoped. The solid waste left over is an environmental hazard; disposing of it is a serious issue that Singapore has mostly handled by dumping it on the small landfill island of Pulau Semakau, but the island will be full by 2035. The incinerators also produce massive amounts of carbon dioxide. And while Singaporean environmental officials claim that the smoke from the incinerators is scrubbed of harmful dust and pollutants, when asked in 2019 if there had been studies on illness rates in neighborhoods near the incinerators, they dodged the question.[29]

But instead of discussing such details, Reinhardt's article lays out a vision of his desired future of limitless energy consumption:

> You could wake up in your house on the beautiful coast of an artificial island off the coast of South America. You're always embarrassed at the cheap synthesized sand whenever guests visit, but people have always needed to sacrifice to afford space for a family. You say goodbye to yours and leave for work. On your commute, you do some work on a new way of making high-temperature superconductors. You're a total dilettante but the combination of fixed-price for infinite compute and the new trend of inefficient but modular technology has created an inventor out of almost everybody. Soon enough, you reach the bottom of the Singaporean space elevator: Cheap space launches, the low cost of rail-gunning raw material into space, and decreased material costs made the whole thing work out economically.* Every time you see that impossibly thin cable stretching up, seemingly into nothingness, it boggles your mind—if that's possible, what else is? You check out the new shipment of longevity drugs, which can only be synthesized in

* This is particularly unusual, because economics isn't the primary barrier to a space elevator. There's no known material that is both light enough and strong enough in long strands to form the cable necessary to create a space elevator on Earth. There is also some reason to think that no such material could exist.

pristine zero-g conditions. Then you scoot off to a last-minute meetup with friends in Tokyo.

As you all enjoy dinner (made from ingredients grown in the same building and picked five minutes before cooking) a material scientist friend of a friend describes the latest in physics simulations. You bask in yet another serendipitous, in-person interaction, grateful for your cross-continental relationships. While you head home, you poke at your superconductor design a bit more. It's a long shot, but it might give you the resources to pull yourself out of the bottom 25 percent, so that your kids can lead an even brighter life than you do. Things are good, you think, but they could be better.

You didn't deal with customs throughout your day because the importance of Westphalian nation states contracted when anybody could be anywhere within two hours. They're still around, but exert an amount of control on where you can live, work, and travel similar to twentieth-century cities.[30]

"Sustainability means perpetual scarcity—in our ability to explore, build, and create," Reinhardt concludes. "I want unbounded possibilities for humanity."[31]

Reinhardt's article appeared in a special issue of *Works in Progress* composed of (almost uniformly positive) responses to a single book, also published by Stripe Press: *Where Is My Flying Car?*, by J. Storrs Hall, the nanotechnology enthusiast and colleague of Eric Drexler. Hall's book reads like a lengthier version of Andreessen's manifesto. It even strikes a similarly aggrieved tone. "If you are a technologist working on some new, clean, abundant form of energy," Hall writes, "you will be attacked, and your invention will be misconstrued and misrepresented by activists, demonized by ignorant journalists, and strangled by regulation. But only if it works."[32] Hall says that there is a vast congregation of "Do-Nots"—mostly environmentalists, bureaucrats, and academics—holding humanity back and keeping "Doers" like engineers and technologists from their rightful place as the moral leaders of humanity. "The Do-Nots favor stagnation and are happy turning our civilization into a collective couch potato," he writes.[33] The solution, he says, is using more energy. Starting around 1800, he claims, per-capita energy consumption grew at a steady rate in the United States all the way up to the 1970s, when it leveled off. "To really reclaim our birthright and an optimistic future," he writes, we must get back on that growth curve, which Hall calls "the heartbeat of our civilization."[34] If we hadn't left that curve, Hall says we would already live in a world that would make Reinhardt's vision look almost pedestrian, with a fusion-powered personal spaceship in every garage, nanotechnology reshaping the plains of West Texas into a duplicate of Yosemite, and a Dyson sphere enclosing the Sun.[35]

Despite the fact that Hall's book is many times longer than Andreessen's manifesto, he doesn't do much better at backing up his claims. The kind of nanotechnology he proposes is the same sort that Kurzweil and Drexler talk about, and suffers from all of the same problems. Hall is also far too credulous about cold fusion, junk science from the 1980s that has long been debunked. He says that "climate change is a hangnail, not a hangman" and waves off its effects as a rounding error in global GDP—which not only is incorrect but also ignores the fact that global GDP is a terrible metric to use for human suffering and environmental damage.[36]* Hall wants a kind of energy-usage singularity, but using more energy

---

* The poorest countries don't contribute much to global GDP, so global GDP doesn't reflect harm done to the residents of those countries, many of which are extremely vulnerable to climate change. In 2009, Nate Silver calculated that you could kill over 40 percent of the population of the world—nearly three billion people—and global GDP would only drop by 5 percent. The fact that this kind of mega-genocide is a rounding error in global GDP is just one of many reasons that GDP isn't a good measure of harm done to humanity, not to mention harm done to other animals, plants, and the rest of the natural world.

won't allow us to invent our way out of basic limits of physics. And the energy-usage-per-capita curve had to stop going up at some point anyhow—it's yet another example of exponential growth. (Humanity's energy usage is still growing steadily; it's only energy usage per capita that's leveled off.) Growth must eventually end; physics demands it, if nothing else. Yet like Andreessen, Altman, and Reinhardt, Hall frames the need for perpetual energy growth in moral terms. "One of the really towering intellectual achievements of the 20th century, ranking with relativity, quantum mechanics, the molecular biology of life, and computing and information theory, was understanding the origins of morality in evolutionary game theory," he writes. "The salient point is that the evolutionary pressures on what we consider moral behavior arise *only in non-zero-sum interactions.* In a dynamic, growing society, people can interact cooperatively and both come out ahead. In a static, no-growth society, pressures toward morality and cooperation vanish: You can only improve your situation by taking from someone else. The zero-sum society is a recipe for evil: it exalts takers while suppressing makers."[37]

There's much to question in this analysis: game theory isn't a set of moral prescriptions; evolutionary psychology has a shaky foundation even at its best. But the most salient rebuttal to the claim that growth is necessary to avoid squabbling and war is that there's been plenty of squabbling and war even while growth has continued unabated. The period of time when human energy usage per capita followed an exponential trend includes both world wars and many others besides. After we left that curve in the 1970s, humanity's energy usage (total, not per capita) and GDP both continued to climb exponentially, yet war and cutthroat competition have continued on in the years since. Strife and competition seem to be substantially, and perhaps entirely, independent of energy usage and economic growth. Even if that weren't the case, we'd eventually need a way to end conflict without exponential growth in energy usage, since maintaining such growth forever is impossible. If Hall were right about everything, and if we even spot him a faster-than-light warp drive, that would just mean we'd be using Dyson-sphere-powered death rays—which Hall actually claims are possible—to wage endless intergalactic war four thousand years from now, when constant growth has led humanity to use all the energy in the universe.

These problems don't faze Hall, and they don't seem to faze Jeff Bezos either. "We want to use a lot of energy. We want to use a lot of energy per capita," he says.[38] "Everybody on this planet is going to want to be a first-world citizen using first-world amounts of energy, and the people who are first-world citizens today using first-world amounts of energy? We're going to want to use even more energy." To do that, Bezos says, we have to leave Earth. We "don't want to face a civilization of stasis, and that is the real issue if we just stay on this planet—that's the long-term issue. . . . Even with improvements in efficiency, you'll still have to ration energy use. And that to me doesn't sound like a very exciting civilization for our grandchildren's grandchildren."[39]

To Hall, there's even more at stake in going to space. "The human society of the future desperately needs a frontier. Without an external challenge, we degenerate into squabbling [and] self-deceiving," Hall writes in *Where Is My Flying Car?* "But the solar system is a foeman worthy of our steel; and, after that, the galaxy is even more so."[40] Reinhardt, meanwhile, is more straightforward. "I like humanity. I think that us going to the stars is good," he tells me. "If somehow I'm still alive in a thousand years, and we're on this trajectory where we're just eating stars [i.e., surrounding them with Dyson spheres] . . . I'll be very happy."[41]

Reinhardt isn't alone in his basic desire to go to space. Elon Musk famously said that "I'd like to die on Mars, just not on impact."[42] The bad news for Reinhardt—and, if we take him at his word, the good news for Musk—is that it's very, very easy to die on Mars.

* * *

In 1972, a somewhat dry book by a team of MIT researchers became a surprise bestseller, ultimately selling millions of copies.[43] The book, *The Limits to Growth*, was a report commissioned by the Club of Rome, a group of scientists, economists, industrialists, and others. The team behind the book had created a model of the world of human activity using a computer—a relatively new idea at the time—incorporating population growth, industrial production, pollution, and limits to natural resources like food and water. The forecasts the computer model returned were bleak. "If the present growth trends in world population, industrialization, pollution, food production, and resource depletion continue unchanged, the limits to growth on this planet will be reached sometime within the next one hundred years," the authors wrote. "The most probable result will be a rather sudden and uncontrollable decline in both population and industrial capacity"—in other words, death and destruction on a global scale.[44]

Limits isn't much discussed today, but the cultural impact at the time was enormous. "Newspapers and magazines revisited the MIT study and its grim conclusions again and again, making the idea of 'limits' a leitmotif for the decade," writes historian and author W. Patrick McCray.[45] It even played a role in setting national and international policy. *Limits* provoked responses from cabinet members and shaped the agenda at UN conferences. Leaders of developing countries worried that the end of growth would lock in a permanent global upper class. China used models similar to the one in *Limits* to develop its one-child policy.[46] Looking back after his loss to Ronald Reagan in the 1980 election, Jimmy Carter said that "dealing with limits" had been the "subliminal theme" of his presidency.[47] But in the 1970s as now, the idea of there being limits to growth—of any kind—wasn't one that everyone was willing to accept. "We have found that technological optimism is the most common and the most dangerous reaction to our findings," wrote the authors of *Limits*. "Faith in technology as the ultimate solution to all problems can thus divert our attention from the most fundamental problem—the problem of growth in a finite system—and prevent us from taking effective action to solve it."[48]

This was an old debate—going back at least as far as Thomas Malthus at the end of the eighteenth century—but *Limits* ignited a particularly passionate and diverse set of responses. Some came from economists like Julian Simon, who later won a bet with Paul Ehrlich (author of another, slightly earlier neo-Malthusian tract, *The Population Bomb*) by correctly predicting that the costs of specific natural resources would decline over the course of the 1980s; some came from conservative commentators, like science fiction author Jerry Pournelle (whose reply to *Limits* made a strong impression on a young Eliezer Yudkowsky).[49]

But it was a response to *Limits* from a Princeton physicist that was arguably the most influential in the long run. Rather than denying the basic premises of *Limits*, Gerard O'Neill wanted to find a way out of the trap it described. Earth is a limited system, so O'Neill proposed finding a way off of Earth entirely. In his 1976 book *The High Frontier*, O'Neill laid out plans for a series of ever-larger space stations. They would mostly take the form of cylinders, rotating rapidly to give the appearance of gravity on the interior of the cylinder. By using launch systems that seemed like a reasonable extrapolation several decades out from then extant technology, O'Neill suggested these space stations could be constructed out of Moon rocks and asteroids, lining the interiors of these spinning cylinders with soil, water, plants, and eventually buildings and people (Figure 5.1a). The smallest of his designs could house thousands of people; the largest would be a suburban metropolis in space, with millions of residents living on an interior surface area larger than Manhattan. The solar system, he reasoned, could easily hold many of these space stations: the raw materials were abundant in the asteroid belt and outer moons, and the energy of the Sun could easily power thousands or millions of these stations. Like today's tech billionaires, O'Neill thought that the resources gained by moving out into space would reduce conflict. Wars on Earth were "battles over limited, non-extendable pieces of land," he wrote, whereas his space stations were "replicable so that no one need feel constrained by a fixed boundary." O'Neill also saw going to space as a way of healing ecosystems on Earth that were being trampled by human activity. Moving most of humanity into space, he thought, would leave behind an "industry-free pastoral Earth" that could serve as a tourist destination for denizens of the cylinders.[50] Ultimately, O'Neill saw this move to space as the only viable option for humanity. There are "three possibilities for a civilization that gets to about our stage," he said. "One is stagnation, one is annihilation, and the third is expansion out into space through space colonies."[51]

To demonstrate the feasibility of these grandiose plans, O'Neill published a series of technical reports and schematics for these space stations, explaining what would be required to build them in detail. He also built a significantly scaled-down version of a "mass driver," the launch system of the future he proposed to move the vast quantities of material needed to build his cylindrical habitats. One of the members of the research team that built that prototype was Eric Drexler, whose voyage into dubious nanotech fantasies was another attempt to escape the limits described by *Limits*; his *Engines of Creation* was indirectly inspired by O'Neill's *High Frontier*.

O'Neill's ideas spread widely over the course of the 1970s and '80s. The L5 Society, named after a stable point in the Earth-Moon system that would be a suitable location for a space station, formed around O'Neill's ideas, promoting them to the public as a viable path to a bigger and better future. Stewart Brand, of the *Whole Earth Catalog*, enthusiastically shared O'Neill's ideas with his large mailing list and set of connections across business and academia. And a certain scientifically minded high school student in Florida was so enthusiastic about O'Neill's ideas that he talked about them to his school newspaper. "The Earth is finite," said an eighteen-year-old Jeff Bezos, "and if the world economy and population is to keep expanding, space is the only way to go."[52] Bezos echoed O'Neill's ideas again in his speech as valedictorian of the class of 1982 at Miami Palmetto Senior High School, where he told his classmates that he wanted "to get all people off the Earth and see it turned into a huge national park," as the *Miami Herald* summarized it at the time.[53] Bezos went on to Princeton, where he attended seminars given by O'Neill.

Today, over thirty years after O'Neill's death, Bezos still sounds like his old professor when he talks about humanity's future in space, right down to his concern about a culture of stagnation if we stay here on Earth. "Do we want stasis and rationing or do we want dynamism and growth?" Bezos asked in 2019. "This is an easy choice. We know what we want."[54] "I would love to see a trillion humans living in the solar system," he said in 2023. "We can easily support a civilization that large with all of the resources in the solar system. . . . The only way to get to that vision is with giant space stations. . . . We will take materials from the moon and from near-Earth objects and from the asteroid belt and so on, and we'll build giant O'Neill style colonies and people will live in those."[55] Like O'Neill, Bezos thinks this will improve Earth—he says that "we will move all heavy industry off of Earth and Earth will be zoned residential and light industry," which, he assures us, will take the Earth back to some prelapsarian state.[56] "500 years ago, pre-industrial age, the natural world was pristine. . . . We have traded some of that pristine beauty for all of these other gifts that we have as an advanced society. And we can have both, but to do that, we have to go to space."[57]

What Bezos doesn't talk about are the problems that have been revealed with O'Neill's cornucopian vision in the intervening decades. The technological advances that O'Neill was hoping would enable the mass driver simply didn't work out (though O'Neill couldn't have known that at the time he was writing *The High Frontier*). Bezos sees Blue Origin, his rocketry company, as his "most important work," because it will enable humanity to live in space.[58] But O'Neill had to postulate the mass driver's existence precisely because he knew that rockets can't move enough mass into orbit to build one of his cylinders. The maximum payload of the most powerful deep-space rocket ever flown as of this writing, NASA's Space Launch System (SLS), is thirty-eight metric tons; it would take over thirteen thousand SLS launches to build the smallest of the habitats that O'Neill designed.[59] (O'Neill had envisioned the mass driver operating on the Moon. Launching rockets from the lower gravity of the Moon would increase their maximum payload, but not by enough to solve this problem—and not all the components of an O'Neill cylinder could come from the Moon. Some things, like people and food, can only come from Earth.) Yet the failure of O'Neill's mass driver is far from the only problem with his plan—or nearly any plan—for humans living in space.

* * *

Elon Musk's plans for Mars do involve more than just dying there. Going to Mars "enables us to backup the biosphere, protecting all life as we know it from a calamity on Earth," he says, like asteroids, nuclear war, or rogue AI.[60] Or, as he put it on Twitter, "We must preserve the light of consciousness by becoming a spacefaring civilization & extending life to other planets."[61] His preferred plan for doing so involves getting people to Mars—at first a few, and then a lot, with the ultimate plan of sending a million people there by 2050.[62] As of this writing, he says he plans to land a SpaceX rocket on Mars by 2029.[63]

While taking Musk seriously is increasingly difficult—it seems likely that he'll say and do many bizarre or hurtful things in the months between the writing and publishing of this book—he still has enormous power and influence, and SpaceX is certainly a serious company, at least for now. It is the sole provider of crewed launches on US soil for NASA (as of 2024), its Starlink system is one of the few options for cell service in many truly remote areas, and future versions of SpaceX's existing Starship launch vehicles could, theoretically, go to Mars. A SpaceX rocket even launched a Tesla out past Mars's orbit in 2018. Musk's timeline for Mars is probably too optimistic—over the years he's given many other dates for boots on Mars and uncrewed landings, and missed them all—but a SpaceX rocket landing on Mars at some point in the next few decades seems like a reasonable possibility.[64] The problem is everything else in Musk's vision. Space—Mars or otherwise—just isn't the place. Nobody's going to boldly go anywhere, not to live out their lives and build families and communities—not now, not soon, and maybe not ever.

Consider Mars. It's fifty-six million kilometers (thirty-five million miles) away at its closest. The most reasonable path there—the route nearly every Mars probe and lander has ever taken—requires about six to nine months in deep space before arriving in orbit around the Red Planet. That's a long time, longer than all but a few humans have ever spent in space, and far longer than anyone has ever spent beyond low Earth orbit. There's a good reason for that: venturing beyond low Earth orbit exposes you to massive amounts of dangerous radiation from the Sun (and a smaller amount from deep space). The Sun is a gigantic nuclear furnace, where hydrogen is built into helium at a temperature of millions of degrees in its core. That blistering nuclear heat eventually makes its way to the surface and atmosphere of the Sun, producing visible light, ultraviolet rays, and other kinds of radiation with even higher energies, like x-rays and fast-moving charged particles. There are also cosmic rays, high-energy radiation produced by violent events beyond our solar system. Here on the surface of the Earth, we're protected from much of this radiation by two mechanisms. The Earth's magnetic field deflects a large amount of the incoming radiation, and our atmosphere absorbs a good deal of the rest before it arrives at the ground. In low Earth orbit, astronauts lack the protection of our atmosphere, and they can see the results: many astronauts have reported seeing occasional bright flashes in the darkness behind their closed eyelids, produced when high-energy radiation slams into their eyes and optic nerves. But such astronauts still have the protection of Earth's magnetic field. Not so if they're on their way to Mars. Astronauts heading into deep space beyond Earth's orbit invariably receive high doses of background radiation. The *Apollo* astronauts each received about 0.4 rads, roughly the equivalent of two head CT scans, in the course of their weeklong trips to the Moon.[65] A trip to Mars would be dozens of times longer than that even if it were just one-way. And if a major solar storm hit the spacecraft on its way out, the crew could be exposed to far greater radiation levels than anything the *Apollo* astronauts experienced. It is possible to shield spacecraft against radiation, but only to a point. Shielding is heavy, which makes it harder to launch the vehicle in the first place. And even heavy shielding can't stop all forms of radiation from getting through over the course of an eighteen-month round-trip journey through deep space.

Bad as it is, radiation is far from the only problem on the journey to Mars. Nine months in close quarters is psychologically taxing even for highly trained astronauts on the International Space Station (ISS)—and they get crew rotations, regular supplies, and real-time communication with the ground. None of that would be possible on a rocket traveling to Mars, which can be up to twenty light-minutes away. Proximity also has other benefits. If anything goes wrong on the ISS, the astronauts can evacuate and be home in a few hours. The astronauts on

Apollo 13 only had to wait an excruciating three-and-a-half days before returning home in their crippled spacecraft. On a Mars mission gone awry, help would be months away—or more than a year. Part of the problem is the distance involved, but the orbital mechanics are also difficult. Unlike trips to low Earth orbit or the Moon, Mars launches are only undertaken at certain times, when the two planets are in the right positions relative to each other. That means no rescue in a reasonable amount of time would be possible for a Mars mission in trouble. Even if nothing goes wrong, there's the dangers of the zero-gravity environment within the spacecraft itself: extended time in zero-g leads to muscle atrophy, bone-density loss, and a variety of other physical ailments. Astronauts coming from extended stays on the ISS have help readjusting to gravity on their return to Earth. But a weakened crew arriving on Mars would have to adapt to gravity without anybody else's help.

Assuming that our intrepid astronauts do make it to Mars in one piece—perhaps with a significantly higher risk of cancer for the rest of their lives, but fine for now—their problems aren't over. Their radiation exposure isn't even over. The surface of Mars receives about as much radiation as nearby points in deep space, because Mars doesn't have a magnetic field and has barely any atmosphere, just 1 percent of Earth's. The best way to shield yourself from radiation on the surface would be to dig underground, using the Martian rocks and dust to absorb the radiation streaming down from above. But that presents a second issue: Martian dust is rich in perchlorates and other toxic chemicals, making it quite poisonous to humans and many other plants and animals of Earth. The good news there is that you'd have to wear an airtight suit even if the dust weren't dangerous. With such low air pressure, astronauts on the Martian surface would have to wear full space suits at all times. Direct exposure to Martian air would boil the saliva off an astronaut's tongue while they asphyxiate; toxic dust would be the least of their concerns. (Although, that toxic dust also has a nasty habit of getting into the Martian air. There are massive dust storms on Mars with alarming regularity, with wind speeds of up to 100 kph [60 mph]. Because the atmosphere is so thin, the storms wouldn't knock astronauts off their feet—*The Martian* is fiction in more ways than one—but they would make it even harder to avoid the dust.) Space suits would also help protect astronauts from the cold climate on Mars, though compared to the other problems we've seen thus far, this isn't so extreme: the average temperatures near the Martian equator typically range from 0°C (32°F) to -70°C (-94°F).[66] So a balmy day on Mars is comparable to a brisk one on Earth, but a brisk day on Mars is as cold as the Antarctic night. And at the Martian poles, it gets far colder than any air temperature ever recorded on Earth, even in Antarctica.

These problems are formidable enough. But if you want to *live* on Mars, rather than just visit for a while, then you have even more problems to handle. Staying on Mars means finding a good source of water, and Mars doesn't really have that. The whole planet is a desert, and its scarce water is contaminated with poisonous dust and other hazardous compounds. Once you obtain that water, you'll need to use it to create a closed ecosystem—probably underground, both for radiation shielding and because that's where much of the water is. That closed ecosystem would need to have plants and microbes (and maybe insects) in order to provide you with the oxygen and food you need to stay alive. In theory, this should work. In practice, nobody has ever done this successfully on a human scale. The highest-profile attempt to create a closed ecosystem with humans in it, Biosphere 2 outside of Tucson, Arizona, had a troubled first mission—oxygen levels dropped steadily over the two years of the experiment—and a second and final mission that ended prematurely. Human factors contributed to the problems there (including the involvement of one Steve Bannon), but it's clear that properly balancing out an entirely isolated ecosystem is a difficult thing to do.[67] It would be even harder on Mars, where there's virtually no oxygen in the air, less than half as much sunlight, and no soil.

This all presumes that it's even possible to get the plants and microbes needed for a closed ecosystem to grow properly in Martian gravity, a third of Earth's. That might be a problem for the humans living there too. We know what extended exposure to zero-g does to humans, and it's not good. We don't know what extended low-g does to humans; ultimately, there's no good way to be sure without conducting highly unethical experiments on humans. And those experiments would look tame compared to the ones you'd need to perform to know whether it's safe to have a family on Mars. We don't know what effects living in a low-g environment would have on pregnant people and kids. It's an open question: it could be fine, it could dramatically shorten their lifespans, or it could kill them. If it's not fine, the only way around it would be to construct an underground centrifuge on Mars large enough for pregnant people to live in until they give birth, and for children to live in until they grow up. A centrifuge for full-g exercise is a good idea, but would it really be reasonable to condemn pregnant people to live inside of it 24-7 for nine months? Would we leave children in there for twenty years? What would *that* do to them? Elton John was right: Mars ain't the kind of place to raise your kids, at least not without information that we can't get without performing truly horrifying experiments on children. Without that information, raising children off-world seems unethically risky.

But merely living on Mars and raising a family there isn't enough to realize Musk's dream. He wants a settlement on Mars to be a backup for humanity. That doesn't just mean a few families, or even a few dozen. If a Mars settlement is going to be a contingency plan for our species in case of a disaster here on Earth, it would need to be fully self-sufficient—as Musk has repeatedly emphasized himself.[68] To do that, you'd need a lot of people. One reason is genetic diversity: in order to prevent dangerous levels of inbreeding and genetic drift among any completely isolated group of people, there would need to be a population of at least a few thousand to start. But self-sufficiency on Mars would require a far larger population than that, because of the technology required to live there. An isolated group of people on Earth can survive with a fairly rudimentary level of technology, as humans did for millennia before the Industrial Revolution and tens of millennia before the development of agriculture. But on Mars, anything but a high-tech society is an instant death sentence. Creating and maintaining the panoply of advanced technology that our society runs on requires a large number of people even here on Earth; Mars wouldn't need less.

So how many people would Mars need for a truly self-sufficient settlement? "One million is actually an absurdly low number of people—far too few to support a modern economy," writes Nobel Prize–winning economist Paul Krugman, in response to Musk's plans for Mars. "Musk's comments immediately called to mind for me a great essay by one of my favorite science fiction writers, Charlie Stross, that posed precisely this question: 'What is the minimum number of people you need in order to maintain (not necessarily to extend) our current level of technological civilization?'" Stross had written on the subject in 2010, concluding that "colonizing Mars might well be practical, but only if we can start out by plonking a hundred million people down there." "If anything, that's on the low side," writes Krugman. Stross agreed—he suggested that the real figure could be as high as a billion people.[69] (Automation won't solve the problem. You still need people to build and maintain the machines, and the economic base needed for a high-tech society would still be large.) Musk's goal of a million people on Mars is unrealistic enough. It's difficult to see how a billion people could live there, and that's ignoring questions of how you'd get that many people off of Earth in the first place. Taking just a million people to Mars would require a rocket launch with a hundred people on it every day for thirty years. And about 1 percent of rocket launches fail, so without serious improvements to the technology, roughly ten thousand people would be killed along the way, sacrificed to the dreams of a billionaire.

Even if you could get a billion people in an underground hyper-megalopolis on Mars, that seems like it could only end in disaster. Keeping a billion people from tampering with key systems, either deliberately or by accident, is an unrealistic goal. That's a lot of hairless apes to keep locked up forever in pressurized tunnels underground. The surface of Mars would be a much better option for a population of that size, if it could be made habitable. Terraforming—finding ways to make the surface of other planets hospitable to unprotected humans—is a staple of science fiction. But the actual science isn't there: we don't know how to do it, though there are many speculative proposals about how it could be done. Musk favors one of these in particular. "Eventually you can transform Mars into an Earthlike planet," Musk said. "Just warm it up. . . . Drop thermonuclear weapons over the poles."[70] Later admitting that was a "little flippant," he elaborated: "What I was really talking about was creating two little suns, two pulsing suns above the North and South Pole of Mars that would warm the poles up enough so that the frozen CO_2 would gasify and densify the atmosphere."[71] That carbon dioxide, along with water vapor released by the heat of the explosions, Musk said, would warm the planet further, releasing still more carbon dioxide and water, eventually leading to a warmer, wetter world with a thick enough atmosphere for humans to survive. Carl Sagan had a more sober proposal for terraforming Mars: plants, which could turn some of the carbon dioxide of the Martian atmosphere into oxygen and warm up the planet with their dark, sunlight-absorbing leaves. Others have proposed bringing in comets and chunks of ice and frozen air from the outer solar system using rockets and smashing them into Mars to deliver more water and atmospheric gases to the planet.

None of these proposals are particularly realistic (though that hasn't stopped Musk from wearing a shirt that says "NUKE MARS").[72]

Nuking the Martian ice caps, releasing all the CO_2 on the planet into the atmosphere, and then vaporizing all of its carbon-rich rocks would still only bring the atmospheric pressure up to a few percent of Earth's at sea level.[73] That's not enough: water would still boil below human body temperature at that air pressure. Plants wouldn't really help with that either. They could indeed convert some of the CO_2 atmosphere to oxygen, but without sufficient air pressure, unprotected humans would still die in minutes on the surface even if there's enough oxygen to breathe. To terraform Mars, more air would need to be brought in from off the planet entirely—hence the proposals to bring in frozen water and gas from elsewhere in the solar system. The technology to do that doesn't currently exist. Even if it could be developed—and it might take decades or more to work it out—it could be very dangerous. If a comet were nudged in its orbit with the intent of hitting Mars, but it missed, its close encounter with Mars would alter its trajectory in unpredictable ways, opening up the possibility that it could hit Earth instead a few years down the line. There wouldn't be much margin for error when maneuvering a comet with such technology—and one comet wouldn't even be close to enough. Thousands of comets would need to be diverted to the Red Planet to have even a chance at giving it an Earthlike atmosphere.

Even if this technology did work out, it wouldn't fix all the problems a human civilization on Mars would face. The dust on the surface would still be poisonous, and people living on Mars would have to figure out how to use that dust to grow food that wouldn't kill them. More generally, they'd have to kick-start an entire closed ecosystem the size of a planet without sending it spinning off into an extreme state, like an algae bloom sucking all the oxygen out of the water. Nobody knows how to do that; working out the science required could easily take centuries, and the answer yielded by science might indicate that such work can't be done more quickly than tens of millennia or more. This would be rather fast on evolutionary and geological timescales but probably more slowly than humans hoping to live on Mars would like. And even if Mars can be made to bloom, there's no solution to its lower mass or lack of a magnetic field. Those two factors would ensure that even a terraformed Mars would still have more radiation at its surface than Earth and less gravity, with all the health problems that come with those.[74]

But let's say that, fifty years from now, Musk's vision has come true, at least to some extent: there are people living on Mars. It's not a billion, it's not even a million, but a few hundred people live in what's intended to be a permanent habitat on Mars. That's probably not impossible. But, given the problems with living on Mars, it wouldn't be pretty. These people would likely be living entirely in tunnels underground; they would rarely, if ever, see the sky. Dangerous excursions to the surface would only be conducted to perform occasional maintenance on machinery. The residents of the small Martian town wouldn't have families—or at least they wouldn't have children—and they would be at the mercy of regular shipments from Earth to replenish their water, food, and technology supplies, meaning that everything would be far more expensive than on a remote island on Earth. All of those shipments would be coming from one supplier: SpaceX, which would have a more complete monopoly than any corporation ever enjoyed in a company town on Earth. Even the air the Mars residents breathe would cost money. Who would truly enjoy living like that? Putting aside the expense involved, it's hard enough to find people willing and able to winter over at the South Pole for eight months. Mars would make Antarctica look like Tahiti.

Despite that, Mars is far and away the most promising option aside from Earth in our solar system. The next most promising is the Moon. The Moon is only three days away, rather than nine months.

That's why we've been able to send people there. But aside from transit time, it's worse than Mars in almost every way. It has half the surface gravity of Mars—about one-sixth of Earth's—and significantly worse temperature extremes: 121°C (250°F) to -133°C (-208°F) at its equator.[75] There's also even more radiation at the surface than there is on Mars. Like Mars, the Moon has no magnetic field, but it also lacks any atmosphere at all, so it gets no protection from radiation. And because the Moon is closer to the Sun than Mars is, more solar radiation hits it. And while Moon dust isn't quite as bad for you as Martian dust, it's not great either. It's made of small slivers of rock that aren't weathered at all, because there's no weather on the Moon. *Apollo 16* astronaut John Young said it was "just like an abrasive," reporting to Houston that it rubbed the text right off all his equipment.[76] And terraforming the Moon is certainly out: its gravity is far too low, and if it was somewhat risky to aim comets at Mars, it would be downright suicidal to aim them at the Moon. A Moon base, like a Mars base, would have to be mostly underground. For all the Moon's faults, its location might make a Moon base somewhat less miserable than a Mars base (though this is of course a low bar to clear). It only takes light a little more than a second to get from the Earth to the Moon, so real-time conversations with friends and family would be possible, unlike on Mars. Goods would be somewhat cheaper on the Moon than on Mars because of the easier transit, and there would be a way to get back to Earth relatively quickly. Otherwise, it would be just as bad as Mars, if not worse. Is there anywhere else in the solar system that wouldn't be like this? Well, maybe.

The surface of Venus is a vision of pure hell, with air pressure as crushing as a kilometer under the ocean and temperatures hotter than molten lead, all under a sky permanently shrouded in clouds of sulfuric acid. But about fifty kilometers (thirty miles) above the surface of the planet, the air pressure is the same as at sea level on Earth, the temperature is usually around 30–60°C (85–140°F), the gravity is only about 10 percent weaker than on Earth, and the thick atmosphere provides good protection from radiation. But Venusian astronauts in an airship at this altitude would still have to deal with concentrated sulfuric acid rain, a near-total lack of oxygen and water, and the constant danger of falling into the crushing infernal depths below.[77] Comparatively, that might not be so bad, but it's still awful—and the conditions on the surface mean that there would be little hope of finding resources on the planet itself to assist with survival. A floating city on Venus would have to get constant resupplies from Earth; without them, everyone in the balloon would starve.

Yet for all their problems, Mars, the Moon, and Venus are easily the best options we have. The rest of the solar system is even more hostile to human life. Mercury is a hotter, more irradiated, and much more distant version of the Moon. Asteroids are airless flecks of rock too small to provide significant gravity unless hollowed out and spun, which would be an engineering project well beyond the current capabilities of our species. The gas giants—Jupiter, Saturn, Uranus, Neptune—offer crushing atmospheres in place of a solid surface. Their hundreds of moons are frozen wastes; even on the closest ones, the Sun shines with about 4 percent of its intensity on Earth. The most promising of them, Titan, has the only other solid surface in the solar system with an atmosphere of the right thickness to allow a person to walk around without a pressure suit, and it has the only known surface lakes outside of Earth. But anyone taking a lakeside stroll on Titan would need to bring their own oxygen, and they'd need to bundle up. The maximum temperature beneath the impenetrable orange haze of Titan's skies is around -180°C (-292°F). The lakes are made of hydrocarbons; water ice on Titan has the consistency of granite. Yet even that is far better than the worlds that wait beyond Neptune. On Kuiper Belt objects like Pluto, the ground is made of frozen air. These diminutive worlds, with surface gravity less than 10 percent of Earth's, slowly amble about the Sun a few times each millennium. Radio signals sent from Earth take hours to arrive.

This is not meant to be an exhaustive list of the problems with all the worlds of our solar system. It's just a sampling of the dangers that would await anyone who ventures there. The solar system is a great place to explore—with robots. For humans, it's wildly inhospitable.

Beyond our solar system, though, there are likely to be many worlds that are much friendlier to human life. Exoplanets, planets orbiting stars other than our own Sun, are one of the great success stories of the last thirty years of astronomy. The first confirmed exoplanet was discovered in 1992; today, there are over five thousand, most of which were found in the last fifteen years. At this point, the scientific consensus is that nearly every star harbors at least one planet. It's only a matter of time before our telescopes spot a likely looking blue-green world orbiting another star, a world of rock and water with an atmosphere something like ours. That will happen, possibly within the next few years, and almost certainly within the lifetime of some of the readers of this book.

But once we find planets like ours, we won't be visiting them. Humans won't be leaving the solar system, for far more reasons than I could list here. Chief among them is the distance involved: the nearest star to our Sun is Proxima Centauri, 4.25 light-years away. Getting a small space probe up to any appreciable fraction of the speed of light is nearly impossible and phenomenally dangerous.[78] For a spaceship carrying humans, it would be even more difficult, and all but certainly lethal to the entire crew. At a third of the speed of light, impact with a single grain of sand would carry a wallop greater than a ton of TNT, ripping the ship apart. Going at more reasonable speeds would mean a journey of centuries or millennia. The astronauts who initially embarked on the journey would die before arriving, as would their children and grandchildren; their distant descendants would arrive at the destination. This sort of "generation ship" is a common feature of science fiction. But generation ships are ethically indefensible. The initial crew would be condemning whole generations of their own descendants to live and die in the confines of the ship, and the generation at the end of the journey would have to try to make a new life on a world they never asked to be sent to in the first place, one that might turn out to be less hospitable than it seemed at a distance. And that's putting aside the practical matter of building a generation ship in the first place: a huge undertaking, requiring vast quantities of material. It would need to be engineered with massive safety redundancies and methods for maintaining careful control of the interior ecosystem, since the whole ship might need to last longer than the present length of recorded history. For a task of this magnitude, it would probably be smart to do a practice run here in our own solar system first—and that's essentially what Jeff Bezos is proposing with his dreams of O'Neill cylinders.

The advantage of giant rotating space stations is that they offer the possibility of creating an Earthlike environment: you can spin them to get artificial gravity on the interior surface, you can make the exterior thick and dense to afford shielding from radiation, and then you can fill them with a breathable, dense atmosphere like the one here on Earth. O'Neill's most ambitious, detailed plans called for cylinders 32 kilometers (20 miles) long and 6.5 kilometers (4 miles) wide, with a surface area comparable to the city of Phoenix, housing several million people. There are insuperable difficulties with getting this much material into space with anything like existing technology, but the problems don't end there. Building a single structure of this size has never even been attempted on Earth; saying that it would dwarf anything previously attempted is a gross understatement. To build this in space would probably require the invention of entirely new disciplines of engineering. It is simply not something we can do now. It's hundreds of years away, if it's even possible at all.

Yet building the structure and shielding of an O'Neill cylinder is arguably the easy part. Once you have the structure, you need to fill it with water, air, plants, animals, and humans, all in a self-contained ecosystem that will be able to run indefinitely without outside help. If Bezos wants a trillion people on space stations while leaving Earth pristine, that means the water and air can't be coming from Earth. They would have to come from spaceborne ice and frozen air, either in comets or the moons of the outer solar system. The station would need at least some soil from Earth to provide the microbes necessary to allow plants to grow; plants would need to be given time to reproduce in the station before humans arrive, to oxygenate the air; insects would need to be released to pollinate the plants. None of this is known to be impossible, strictly speaking, but we don't actually know if it's possible either. What we do know is that it would be an enormous task to build one of these stations. A comparison of O'Neill's plans with extant space stations gives a good sense of just how far beyond our current capabilities such a station would be: compare NASA's depiction of the interior of an O'Neill cylinder in Figure 5.1(a) with the interior of the ISS, far and away the largest space station ever actually built, in Figure 5.1(b). But Bezos doesn't just want to build one. If each cylinder houses five million people, then realizing his plan of a trillion people in space would involve building two hundred thousand of these behemoth stations. Hundreds or thousands of comets would be needed just for the water and air on these stations.[79] Building the actual structure of the cylinders themselves would far exceed that, and would likely involve disassembling a large asteroid for its mass—a task completely beyond any technology humanity currently knows how to build.

In the unlikely event that Bezos and his followers succeed in this improbable quest, they would end up with a fleet of hundreds of thousands of space stations, each filled with millions of people who can't easily leave—not a company town, but a company civilization. Sticking

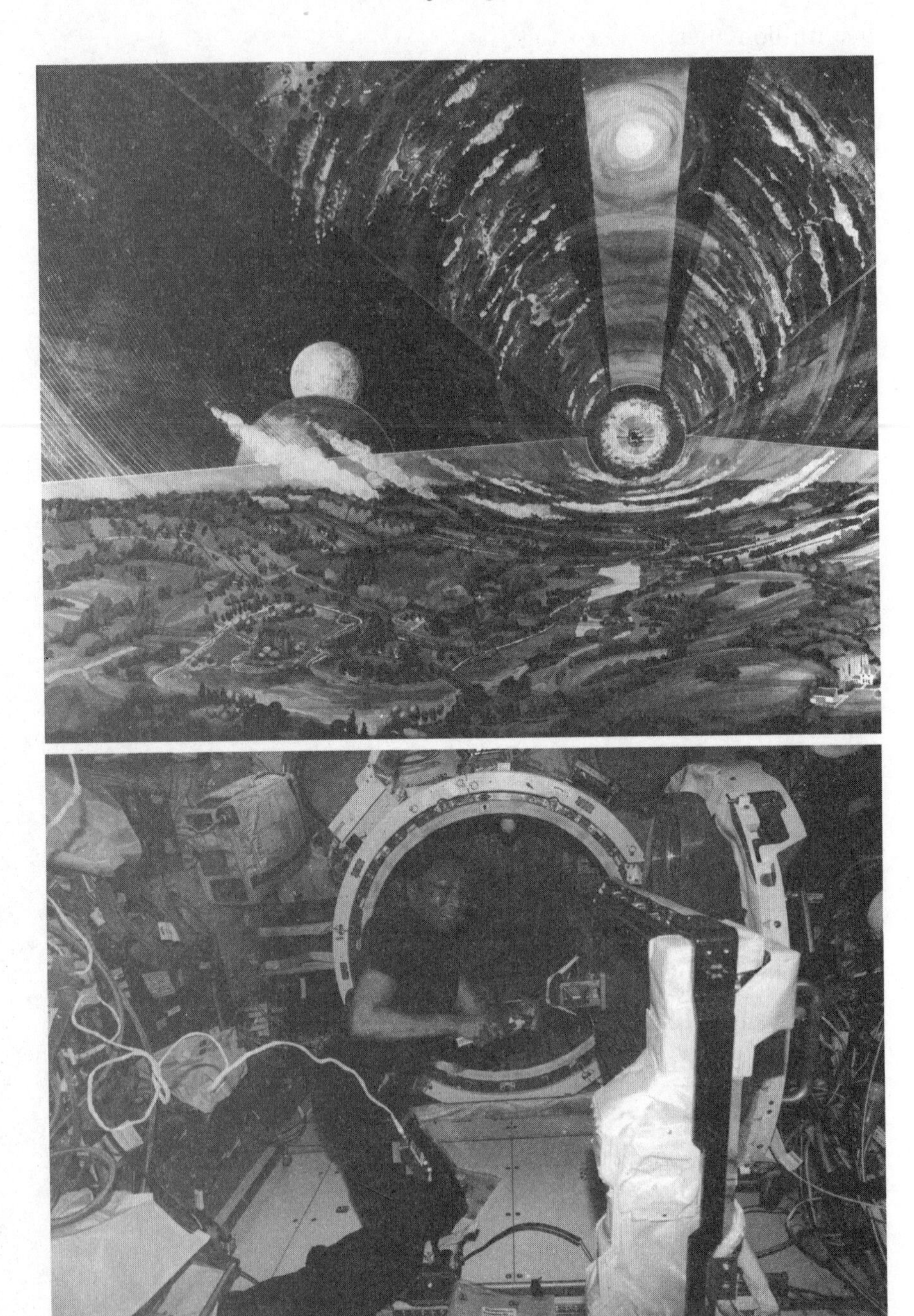

Figure 5.1: (a, top) A NASA concept illustration of the interior of an O'Neill cylinder. (b, bottom) NASA astronaut Jeanette Epps aboard the International Space Station, the largest space station ever built.

five million apes in a can together—even a big can—will necessarily lead to politics of some kind. And with a trillion humans in space, it's only a matter of time before something goes horribly wrong. A healthy adult dies from direct exposure to the vacuum of space within about three minutes, if they exhale first to keep their lungs from ripping apart. In Bezos's future, all it will take is one malfunction, one accident, one act of lunacy or malice to kill millions—or billions—in less time than it takes to eat an apple.

There is one final option, though. There's a spot in the solar system that has the gravity, the air, and the water we need. It has a strong magnetic field that deflects radiation, a temperature range we can handle with existing technology, and a twenty-four-hour day too. There's even an ecosystem already in place, one that grows food that humans can eat. And best of all, we don't have to worry about how to get a self-sustaining human population going, because there are already eight billion people living there.

* * *

Despite all of the problems, it's still possible that humans—maybe even large numbers of humans—might live their entire lives in space someday. The point is merely that we might *not*, because it's hard not to die in space. And if people do eventually build a self-contained civilization in space, it's probably going to take a long time, millennia or more, for that to happen.

Just because something is difficult and time-consuming doesn't mean it's not worth trying to do. John F. Kennedy famously said that the United States should send astronauts to the Moon precisely because it was hard. That wasn't the real reason the *Apollo* program happened—that was about competition with the USSR during the Cold War. Compared to that, Musk's desire to back up humanity on Mars in the event of a disaster on Earth sounds noble. Other tech billionaires agree. "[Musk] wants to go to Mars, to back up humanity," says Google cofounder Larry Page. "That's a worthy goal . . . and it's philanthropical."[80]

But Page is wrong. Putting aside the inherent problems with billionaire philanthropy, Musk's plan of backing up humanity on Mars is still a bad one, precisely because Mars is so awful. The idea of a backup ostensibly comes from a fear of humanity going extinct in a planetwide catastrophe, like global warming, nuclear war, or a large asteroid strike. Of those options, big asteroids are probably the most destructive. They're certainly the most dramatic. The single worst day in the history of complex life on Earth happened sixty-six million years ago, when a rock as wide as Brooklyn and taller than Mount Everest slammed into the Yucatán Peninsula one hundred times faster than a jumbo jet. That asteroid punched a hole nearly all the way through the crust of the Earth to the mantle, launching countless tons of rock deep into space—fragments of dinosaur bone from this impact are likely still on the Moon, even now.[81] Much of the rocks ejected from the impact fell back to Earth within hours, burning across the sky. The heat of their reentry turned the air red-hot in many places around the globe, igniting widespread wildfires. In some places, the surface air was hotter than an oven set to broil. Creatures that survived the day faced more horrors. Billions of tons of sulfur dioxide flung into the atmosphere from the impact blotted out the Sun for years, dropping global temperatures and killing plants and photosynthetic microbes all around the world, on land and in the ocean. With the bottom of the food chain cut out from under them, most creatures starved. Ultimately, 80 percent of animal species went extinct, famously including all non-avian dinosaurs.[82]

But "when the asteroid hit 66 million years ago it was a nicer day than today on Mars—otherwise no animal life would have survived," says Peter Brannen, author of *The Ends of the World*, a survey of mass extinctions on Earth. "So we better learn to live here. This is it."[83] Our existence is proof Brannen is right: despite the horrors of that day, mammals survived, as did birds, fish, and many other vertebrates. No mammal could survive unprotected for more than a few minutes on Mars. "We'll never trash this place so bad with nukes or climate change that somewhere else is a better option," Brannen concludes.[84] Or, as Martin Rees, cosmologist and Astronomer Royal, puts it: "It's a dangerous delusion to think that space offers an escape from Earth's problems. We've got to solve these problems here. Coping with climate change may seem daunting, but it's a doddle compared to terraforming Mars. No place in our solar system offers an environment even as clement as the Antarctic or the top of Everest. There's no 'Planet B' for ordinary risk-averse people."[85] If you knew, right now, that another asteroid was bearing down on Earth like the one that hit sixty-six million years ago, and if you were offered the choice of staying here or going to a base on Mars like the one Musk wants to build, staying here would be the far better bet. Everyone here would probably die, but without supply missions from Earth, everyone on Mars would *definitely* die.

Proposals to terraform Mars are particularly bizarre in light of global warming here on Earth. Terraforming Mars, with all its myriad challenges, is unequivocally more difficult than solving global warming on Earth, which merely requires that we stop pumping so much carbon dioxide and other greenhouse gases into the atmosphere (and, hopefully, find good ways of removing some of what we've already put there). Yet humanity has proven fairly bad at addressing global warming to date. Many of the same people who want to terraform Mars would rather find ways around the need to stop burning fossil fuels by trying to "terraform" Earth instead through geoengineering. Such plans have been floated for decades; they've generally been rejected for the very good reason that their full effects are difficult to predict, with a significant chance of unintended consequences that could be worse than those of global warming in the first place.

Even if it is possible to terraform Mars, developing the technology to do so could make it more likely that humanity will destroy itself, not less. The twentieth century demonstrated that rival nuclear superpowers would readily engage in a space race. Imagine space drones that can bring comets in from the outer solar system and direct them to hit the surface of Mars. Now imagine that your country doesn't have them, and its greatest rival does. How do you think the leaders of your country would feel about that? An arms race would develop, and it wouldn't take advanced terraforming technology to set it off—nearly any of the technological advances needed to live and work in space would do. A Moon base that could launch the raw materials for an O'Neill cylinder into space could also launch massive Moon rocks at any location on Earth. Those rocks, falling down from the Moon's orbit to the surface of the Earth, would hit the atmosphere traveling at about ten kilometers per second (six miles per second), half as fast as the asteroid that killed the dinosaurs—easily powerful enough to cause death and destruction on a scale comparable to nuclear weapons.[86]

Nor is an arms race the only kind of international crisis that could be precipitated by an attempt to colonize space. In 2020, Starlink, SpaceX's satellite internet service, made headlines with a clause buried in their user agreement that stated, in part, that "the parties recognize Mars as a free planet and that no Earth-based government has authority or sovereignty over Martian activities. Accordingly, Disputes will be settled through self-governing principles, established in good faith, at the time of Martian settlement." But there are existing international treaties governing territory in space that can't be abrogated with an end-user license agreement. That agreement "made me laugh," said Bleddyn Bowen, a lecturer in international relations and space policy expert at the University of Leicester. "The language in that clause violates the 1967 Outer Space Treaty, and especially Articles VI and VIII."[87] David Koplow, a law professor at Georgetown University, said,

"I have no idea what they might mean by 'free planet' or 'self-governing principles'—from the standpoint of international law, that's just gibberish."[88] The Outer Space Treaty states that the Moon, Mars, and every other celestial body belongs to humanity—and that space travelers are still under the jurisdiction of their home nations. Right now these legalities are largely abstract, but they could become awfully concrete in the event of competing Moon or Mars bases from different countries. If one country does something that they think is legal in space, but other countries disagree, this could lead to an unstable international situation and ultimately to war between nations with advanced capabilities in space.[89]

So building a large city on Mars probably isn't possible in the first place, it wouldn't help save humanity in the event of a disaster here on Earth, it could easily make humanity less safe rather than more so, and it's legally questionable. Why does Musk want to do it? "A lot of times when you hear the narrative about creating a backup for humanity, it's actually not climate change that most of those people bring up. What they often bring up is asteroid strikes. And I always have found that super interesting," says astronomer Lucianne Walkowicz. "Because think about what an asteroid strike is. It's something that comes out of the blue, that you have absolutely no way of preventing, that you didn't know was going to happen. And I think the reason that you don't hear climate change in the mouths of people who [say] we need to create a backup for humanity—maybe they will talk about it in another forum, but they don't talk about those two things together—is because climate change implies human responsibility. Climate change implies actions that they are responsible for." Rather than face that responsibility, Walkowicz suggests, billionaires are searching for a way to avoid it. "I absolutely despise this idea of backing up humanity. . . . [Billionaires] have essentially the most means and the most power to do something about the problems that we actually face, whether they be climate change or income inequality," they say. "This idea of backing up humanity is about getting out of responsibility by making it seem that we have this Get Out of Jail Free card."[90]

This is painfully apparent in the way Jeff Bezos talks about geniuses of the future. He dreams of one thousand Mozarts and one thousand Einsteins among his trillion humans living in space. But he's neglecting the potential Einsteins and Mozarts that are living and dying in poverty right now. If Bezos really wants a dynamic civilization, he could invest in solving the problems that we actually face. He could tackle the enormous levels of wealth inequality in the world today. He could put all his resources into fighting global warming, which will also have unequal impacts around the world. Instead, he's going to space—and he thinks that's the best and most important thing he can do with his money.

This is the promise of transcendence offered by technological salvation. Upton Sinclair said that "it is difficult to get a man to understand something when his salary depends upon his not understanding it."[91] But for Bezos and Musk, there's more than a paycheck at stake. These dreams of space, earnestly held in spite of all the evidence against them, allow these billionaires to avoid responsibility for any problems here on Earth and instead claim that nothing could be more important than what they want to do. No real crisis can get in the way of the imagined mission. During the height of the COVID-19 pandemic, Musk pushed Tesla to violate shelter-in-place orders, and Bezos's Blue Origin pressured workers in Washington State to travel to rural Texas to test launch a rocket for space tourists.[92] "There will be no Mars if we let them take our freedom away," Musk tweeted in May 2020, regarding pandemic precautions.[93] Here Musk is identifying his plans with Mars itself. Criticism of Musk's power fantasies becomes criticism of the culturally powerful idea of space, an idea he returned to in a tweet the next year:

those who attack space
maybe don't realize that
space represents hope
for so many people[94]

But hope for whom, exactly? "What would it actually look like to backup humanity? Who's humanity? Who's included in the backup?" asks Walkowicz. "When we look at the exclusionary nature of astronaut selection . . . if we were to do something like that tomorrow without unpacking a lot of the ways in which we've constructed the idea of who is worthy of even going to space, let alone being the person who gets on the ark for humanity—I think you have a very close tie to eugenics there," they say. "Already, we see that disabled people are disqualified [from being astronauts]. Historically, women have been disqualified; people of color have been disqualified. And that's been really a running theme throughout the astronaut program. Even though the astronaut program is more diverse now, it is still fundamentally an exclusionary system. Some of that might have to do with the demands of space, and some of it is just ideas that people have about whose body is worthy."[95]

This exclusionary subtext also shows up in the most common term these billionaires—and many, many others—use for sending large numbers of people to live off Earth: space *colonization*. The word colonization "is tied to a history that fundamentally has benefited people of European descent who are in the Americas," says Walkowicz. "Colonization is still an everyday part of life for a lot of people, and so when we use 'colonization' to talk about space in not even a neutral way—when colonization is spoken about in the context of space, it's in this shiny future—that erases the history that exists here on Earth. . . . When we use it in these neutral or even aspirational ways, we are really actually perpetuating a harm that eschews that responsibility."[96] The late physicist Stephen Hawking gave a speech in 2017 that was a good example of this kind of thinking. "The Earth is becoming too small for us, our physical resources are being drained at an alarming rate," he said. "When we have reached similar crises in our history there has usually been somewhere else to colonize. Columbus did it in 1492 when he discovered the new world. But now there is no new world. No utopia around the corner. We are running out of space and the only places to go to are other worlds."[97] Who is "we" in that paragraph? It's certainly not the Native Americans, who got on quite well for millennia before Columbus and other European colonists invaded their land and killed off 90 percent of their population. What are Native Americans to infer about their place in plans for space that use the word "colonization?"

The rapacious logic of colonialism pervades the dreams of technological salvation in space. There's Kurzweil and Bostrom asserting confidently that aliens probably don't exist, so the resources of the universe are ours for the taking. There are the plans for the exploitation and destruction of nature on a cosmic scale. There are the billionaires planning out visions for our manifest destiny in space, regardless of what we already know and what else—or who else—we might find there. Even in the event that we never encounter alien life of any kind, the logic of colonialism has a history of violence that could play itself out again among humans in space, this time with enormously more dangerous technologies at hand.

The fact that Musk and Bezos—not to mention Hall and Reinhardt—continue to use the language of space colonization in the face of these issues is just one instance of a bigger problem: there's a peculiar blindness to history here, especially the historical perspective of people who aren't white men. "Whatever the proximate causes of the cultural shift in the 1960s and '70s, the one incontrovertible fact is that it happened," Hall writes in *Where Is My Flying Car?* "The culture of trust and 'same-boat spirit' of the '30s, '40s, and '50s evaporated into the culture wars of the '60s and '70s."[98] Claiming that there were no culture wars in the United States from 1930 to 1960 betrays a shocking ignorance of history on Hall's part. There was the massive social unrest of the Great Depression, with calls for revolution of all kinds; there was the civil rights movement, which was highly active in the 1950s and arguably started in the 1940s; there were the seeds of second wave feminism that sprouted after World War II; there was the Dixiecrat walkout from the Democratic Party in 1948, motivated purely by racism—the list goes on. But when Hall looks at that era, he doesn't see any of that. He just sees the perspective of comfortably well-off white men, who certainly had a lock on power in the United States throughout that period (and well before). Similarly, when the tech billionaires and their hangers-on look at the idea of colonialism, they don't see genocide and exploitation. They see a frontier where they can escape regulation, along with any other consequences for their actions here on Earth.

* * *

Apparently, expecting tech billionaires like Musk, Bezos, and Andreessen to pay attention to facts and make reasoned arguments for their beliefs is asking too much of them. Andreessen's manifesto certainly isn't running on logic—it's running on pure vibes. "To paraphrase a manifesto of a different time and place: 'Beauty exists only in struggle. There is no masterpiece that has not an aggressive character. Technology must be a violent assault on the forces of the unknown, to force them to bow before man.'" Andreessen continues in that vein for a while, working himself up. "We are not victims, we are *conquerors*. . . . We are not primitives, cowering in fear of the lightning bolt. We are the apex predator; the lightning works for us. We believe in *greatness* . . . ambition, aggression, persistence, relentlessness—*strength*."[99]

The "manifesto of a different time and place" that Andreessen is referencing is the *Futurist Manifesto*, written in 1909 by Filippo Tommaso

Marinetti, an Italian poet and art theorist. (It's a very close paraphrase; all Andreessen did was replace "poetry" with "technology.") "We want to glorify war—the only cure for the world—militarism, patriotism, the destructive gesture of the anarchists, the beautiful ideas which kill, and contempt for woman," Marinetti wrote. "We want to demolish museums and libraries, fight morality, feminism, and all opportunist and utilitarian cowardice."[100] The connection to Marinetti is helpful for identifying the vibe Andreessen is seeking to create here, with all his talk of aggression, conquering, and apex predators. Ten years after writing the *Futurist Manifesto*, Marinetti coauthored the *Fascist Manifesto*, and shortly thereafter he became an early and enthusiastic supporter of Benito Mussolini. Marinetti maintained his support for Mussolini and fascism until his death in 1944.

Andreessen insists that techno-optimism isn't a political position, and that it is not necessarily left- or right-wing (though socialism and communism are on his list of enemies). Yet Andreessen not only quotes Marinetti in the manifesto but also lists him among the "Patron Saints of Techo-Optimism." Nor is Marinetti the only fascist on Andreessen's list. He also includes Nick Land, a neoreactionary philosopher widely read among the alt-right followers of Curtis Yarvin. (The effective accelerationists also quote Land in their inaugural post.)[101] Andreessen put Frederick Jackson Turner on the list too. Turner was a late-nineteenth- and early twentieth-century American historian who articulated and advocated for the "frontier thesis," the idea that the frontier was essential to the vitality of American culture. This colonialist rhetoric is echoed in Andreessen's manifesto, in Hall's talk of the solar system as a necessary challenge for our civilization, and in O'Neill's visions of the high frontier. In 2016 Andreessen said that "anti-colonialism has been economically catastrophic for the Indian people for decades," apparently in response to the Indian government shutting down an initiative from Facebook that was providing free access to a highly restricted version of the internet. It's unclear how that action outweighs the horrors of the British occupation of India for Andreessen—he didn't elaborate, though he later attempted an apology and claimed he was "100% opposed to colonialism"—but again, this is all about vibes.[102]

Those vibes—vibes of fascism, authoritarianism, and colonialism—are fundamentally about creating a fantasy of control, the ultimate drug offered by the ideology of technological salvation. And once again, the distance between Andreessen's effective accelerationist camp and that of the effective altruists and rationalists is vanishingly small. The effective altruists and rationalists also want control, control over the superintelligent AI that will set humanity on the best path to the future of highest value. (They have also deployed similar racist and authoritarian rhetoric along the way.) Andreessen seems to want that too; he just differs about how to get there, what constitutes value, and who should be in charge of telling the superintelligent AI what to do as it determines our future. If you want a picture of that future, imagine a billionaire's digital boot stamping on a human face—forever.

With that dream of eternal universal control comes another component of technological salvation: using that perfect control to transcend all limits. Here, Andreessen and the techno-optimists actually manage to outdo the effective altruists. The effective altruists are certainly interested in using technology to transcend many limits, but they recognize that at least some hard limits exist. Bostrom wants to get out to space precisely because he knows that the amount of low-entropy energy is limited and eventually the heat death will come; MacAskill explicitly argues for the implausibility of endless economic growth, even in a universe-spanning empire, because he knows that relentless exponential growth would eventually demand squeezing a planet's worth of GDP from every subatomic particle in the cosmos. Even Reinhardt is willing to concede that growth has to end at some point, given the finite nature of the universe. But Andreessen really believes that growth can—and must—continue forever, driven by technology. Tellingly, the final entry on Andreessen's enemies list is "the limits of growth."[103] This is almost certainly a reference to *The Limits to Growth*, but Andreessen makes it clear in the rest of his manifesto that he does not believe that any limits to growth exist at all.

The claim that limits to growth exist—not necessarily the specific limits forecast by the Club of Rome, but that some limit exists—is so basic that countering it seems like the purest denial, a toddler's howl of rage at being told no. And there's surely a great deal of resentment coursing through the rhetoric of these billionaires and their pet intellectuals. Resentment is the underside to their false bravado, dripping off nearly every sentence of Andreessen's manifesto. More than anything, he seems to resent expertise. "Our enemy is the ivory tower, the know-it-all credentialed expert worldview, indulging in abstract theories, luxury beliefs, social engineering, disconnected from the real world, delusional, unelected, and unaccountable," he writes. "We believe in the *actual* Scientific Method and enlightenment values of free discourse and challenging the authority of experts." This is reflexive contrarianism as a virtue, trying to claim the mantle of science and the Enlightenment while saying that his ignorance is just as good as someone else's knowledge.[104] (Andreessen's professed belief in science is especially rich given that he includes George Gilder, a creationist, on his list of "Patron Saints of Techno-Optimism.") And just in case it wasn't clear, Andreessen wants everyone to know that he isn't resentful—not in the least. He is careful to reassure his readers that "we believe in an absolute rejection of resentment."[105]

A powerful billionaire criticizing experts as disconnected and unaccountable is an astonishing feat of projection. Andreessen is portraying himself as heroic, the great man against the world, fighting an entrenched establishment. In reality, he is the establishment. As the writer Jon Bois says, "The thing about Goliaths is that they always want to be David."[106] Rather than facing the true limits on his impossible dreams, imposed by simple facts about this world that no human made, Andreessen has apparently decided that he's just a plucky underdog fighting the good fight against the rest of us. Seeing through the trick he's played on himself would require confronting the ultimate limit of human life, one that no amount of money can eliminate. It's the same one the rationalists, the effective altruists, and Kurzweil are all scrambling to avoid. Lurking underneath all of the dreams and desires and resentment of the tech billionaires lies a fear of death, the final loss of control. Hall, who is in his seventies, is rather open about this. "If you would have a project that had, say, a billion dollars a year. . . . I think you could probably get [Drexler-style nanotech] within a decade," he tells me. "If we don't do it, I'm going to die."[107]

6

WHERE NO ONE HAS GONE BEFORE

When I was a kid, I thought *Star Trek* was a documentary about the future. I didn't literally think that Captain Picard and Data and Geordi and the rest of the gang would be hanging out on the holodeck in the twenty-fourth century, but I did think that this was the future that smart adults had worked out as the best one, that this was what we were going to do, or at least what we should do. We'd go to space, we'd seek out new life and new civilizations, and we'd do a lot of science. I was six, and that sounded pretty good to me.

But my love for the show went deeper than that. *Star Trek* didn't just show a bright future; it showed a group of people being thoughtful and kind to one another, working together to solve problems, taking care of their friends. I didn't have a lot of friends at that age, and I had some problems at home too. I desperately wanted to be on the *Enterprise*, because I wanted to spend time with those characters, who had come to seem like my friends. I wanted to hang out with them. In lieu of being there, I raided my dad's collection of science fiction novels and short stories—mostly from his childhood in the 1950s and '60s—and tore through the space and physics section of the library at my elementary school before doing the same at my hometown's small public library, a short walk from my house.

The town library also had a collection of VHS tapes—one that seemed vast to me at the time, but that couldn't have comprised more than a few dozen movies and PBS specials. Kids weren't allowed to check them out on their own, but one day when I was about eight, my mom went up to the library after she got home from work and brought back a small pile of tapes, all emblazoned with the same title that had been cut out of the original box and Scotch-taped onto the blue marbled plastic of the library's boxes: *Cosmos*. I watched Carl Sagan with uncut fervor and awe. Here was someone not only talking about the grand sweep of the scientific picture of the universe but also making a case that we could go to space, that we could save the world, that we could build a better future together. To me, at that age, it sounded like he was saying what I'd been wanting to hear: we could make *Star Trek* real. Sagan, of course, never actually said that, though he was pretty optimistic about deep space travel and terraforming other planets, and that's all it took for me. I knew what the future held. It was a bright promise, and I held it close.

That promise, in turn, made certain things easy. There were lots of books at the library, but now I knew that I didn't have to pay attention to most of them. Anything that wasn't science—really, anything that wasn't physics and astronomy—was just messy details. None of it mattered, not really. The only thing that was actually important was going to space to save the world.

As I got older, my interests broadened. I wanted to know how we knew the things we did, and that led me to history and philosophy of science. I wanted to understand why the adults had created such a strange world, and why nobody seemed to be doing anything about global warming, and that led me to political science and more history. I read more adventurously, never leaving science fiction behind but taking my chances with other kinds of books more often, and reading more (and less!) recent entries in the genre than the "Golden Age" stories I'd been raised on. But through all that, I never really lost the faith that *Star Trek* had given me. The future, I knew, ultimately lay in space, and going there would solve many—maybe even all—of the problems here on Earth.

I believed that for a long, long time.

* * *

There's an odd little throwaway line in Benjamin Reinhardt's vision of an "ergophilic" future, one that didn't even register properly the first time I read it: he talks about "longevity drugs" that "can only be synthesized in pristine zero-g conditions" out in space. While it's not impossible that a zero-gravity environment could help with drug manufacturing, I suspect that this wasn't what Reinhardt had in mind, or at least wasn't the only thing he had in mind. Consciously or unconsciously, he was depicting the same old dream: going to space and living forever.

But why does it so often seem that the idea is to go to space *in order* to live forever? What does space have to do with immortality? The interdisciplinary scholar of AI Kate Crawford suggests an answer:

> Space has become the ultimate imperial ambition, symbolizing an escape from the limits of Earth, bodies, and regulation. It is perhaps no surprise that many of the Silicon Valley tech elite are invested in the vision of abandoning the planet. Space colonization fits well alongside the other fantasies of life-extension dieting, blood transfusions from teenagers, brain-uploading to the

> cloud, and vitamins for immortality. Blue Origin's high-gloss advertising is part of this dark utopianism. It is a whispered summons to become the Übermensch, to exceed all boundaries: biological, social, ethical, and ecological. But underneath, these visions of brave new worlds seem driven most of all by fear: fear of death—individually and collectively—and fear that time is truly running out.[1]

This—the desire to transcend all limits and escape the terror of death—is surely part of the answer. But the linkage of space with immortality predates the tech billionaires and the modern transhumanist movement. In the 1970s, Timothy Leary was asking people if they "would like to live in space and live forever." Inspired in part by O'Neill, he promoted "Space Migration, Intelligence Increase, and Life Extension," anticipating the Extropians by over a decade.[2] The L5 Society's newsletters frequently featured ads related to life extension and cryonics.[3] Even cryonics itself was originally tied to space: Robert Ettinger, the physics instructor who started the modern cryonics movement with his 1964 book *The Prospect of Immortality*, was inspired by "The Jameson Satellite," a short story by Neil R. Jones in the July 1931 issue of the science fiction magazine *Amazing Stories*. In the story, a professor (the titular Jameson) is launched into space after his death, where he orbits Earth, frozen solid. After millions of years, he is revived and placed into a mechanical body, becoming immortal.[4]

So perhaps there's something simpler, or at least older, at work here, alongside the ideology of technological salvation. Maybe it's just that immortality comes from heaven. "Today artificial intelligence and information technologies have absorbed many of the questions that were once taken up by theologians and philosophers: the mind's relationship to the body, the question of free will, the possibility of immortality," writes essayist Meghan O'Gieblyn, author of *God Human Animal Machine*. "All the eternal questions have become engineering problems."[5] O'Gieblyn used to believe in a form of transhumanism; before that, like many other transhumanists, she had been a fundamentalist Christian, studying theology at Bible college. "I had believed since childhood that earthly life was an arc bending toward a point of final redemption, to the moment when Christ would return, the dead would rise, and the entire earth would be restored to its original perfection. This was not in any sense metaphorical. . . . For most of my life I had believed that I would live to see the coming of this new age; that my body would be transformed, made immortal, and I would ascend into the clouds to spend eternity with God."[6]

O'Gieblyn stopped believing in God—"without that narrative, my life lost its mooring"—and several years later encountered *The Age of Spiritual Machines*, by Ray Kurzweil. For a time, she was obsessed. "What makes transhumanism so compelling is that it promises to restore through science the transcendent—and essentially religious—hopes that science itself obliterated," she writes.[7] "My interest in Kurzweil and other technological prophets was a kind of transference. It allowed me to continue obsessing about the theological problems I'd struggled with in Bible school, and was in the end an expression of my sublimated longing for the religious promises I'd abandoned." Eventually, she stopped believing in Kurzweil's predictions too. But she was left with a persistent question. During her time as a transhumanist, she had noticed "strange parallels between transhumanism and Christian prophecies. Each time I returned to Kurzweil, Bostrom, and other futurist thinkers, I was overcome with the [conviction] that the resonances between the two ideologies could not possibly be coincidental."[8] Reading more broadly, she discovered "that there had existed across the centuries a long tradition of Christians who believed that the Resurrection could be accomplished through science and technology."[9]

One of these Christians was Nikolai Fedorov, a late-nineteenth-century Russian philosopher and librarian. "The awaited day [of resurrection] is the hope of all ages and peoples, awaited from time immemorial," Fedorov wrote. "This day will be divine, awesome, but not miraculous, for *resurrection will be a task not of miracle but of knowledge and common labor.*"[10] In order to resurrect the dead, Fedorov maintained, humanity would first have to obtain total command over the natural world—"the regulation of nature by human reason and will," as he wrote to a friend—in order to eliminate all causes of death and decay.[11] With mastery of nature, he thought, advances in science could eliminate death for those currently living. Then we would move on to resurrecting the dead. Fedorov claimed this would require going into space to recapture some of the physical material of dead bodies, which he believed had escaped there. Once that had been done, Fedorov was confident that advances in science would make it possible to resurrect all of humanity's dead and fill the cosmos with everyone who had ever lived, making the world "as it ought to be."[12]

Fedorov's philosophy, which came to be known as cosmism, was warmly received by his contemporaries Fyodor Dostoyevsky and Leo Tolstoy. But Fedorov's most lasting mark on the world was through his tutelage of the physicist Konstantin Tsiolkovsky. In 1874, the sixteen-year-old Tsiolkovsky first visited the Moscow library where Fedorov worked.[13] Tsiolkovsky was soon meeting Fedorov every day, talking with him for hours.[14] "Understanding my inclination toward mathematics, physics, and in part, chemistry, he [Fedorov] selected literature for me and directed my self-education," Tsiolkovsky said later. "It is no exaggeration to say that for me he took the place of university professors, with whom I had no association."[15] There are conflicting accounts of whether Tsiolkovsky and Fedorov actually discussed space exploration directly, but either way, Tsiolkovsky certainly read Fedorov's work later on, and the schooling the younger man received from the older set him on a course for the stars.[16] In 1903, the year Fedorov died, Tsiolkovsky published his monograph "Exploration of the

Universe with Reaction Machines," one of the first scientific papers to lay out the basic physics of rockets.[17] Like his mentor, Tsiolkovsky was convinced that humans needed to travel to the stars. "Earth is the cradle of humanity, but one cannot live in the cradle forever," he wrote.[18]

That quote—which is on Tsiolkovsky's tombstone—is sometimes translated from the original Russian as "This planet is the cradle of human mind, but one cannot spend all one's life in a cradle."[19] This translation suggests some of the philosophical similarities between Tsiolkovsky and Vladimir Vernadsky, a contemporaneous Russian scientist and cosmist. Vernadsky was a geochemist who popularized and extended the idea of a "biosphere." He also developed the idea of a "noosphere," a global network of human thought and knowledge emerging atop the biosphere. Vernadsky's idea of the noosphere influenced (and was influenced by) Pierre Teilhard de Chardin, a French paleontologist and Jesuit priest.[20] Teilhard combined the idea of the noosphere with evolutionary theory and his Catholic faith to arrive at surprising—and surprisingly familiar—conclusions. "Teilhard believed that evolution was not only ongoing but was developing at an exponential rate," O'Gieblyn writes:

> Humans, through their use of tools and mechanization, were now in a position to direct the course of their own evolution. The invention of radio, television, and other forms of mass communication had created complex global networks that facilitated more intricate and intimate connections between individual minds. . . . Teilhard set out a vision for how these technological connections, which he called "the noosphere," would eventually lead to a dramatic spiritual transformation. In the future the network of human machines would give way to an "'etherised' universal consciousness" that would span the entire circumference of the globe. Once this synthesis of

> human thought reached its apex, it would initiate an intelligence explosion—he called this the Omega Point—that would enable humanity to "break through the material framework of Time and Space" and merge with the divine. . . . The resonances between this vision and Kurzweil's prophecies are uncanny. And yet Teilhard believed that this was how the biblical Resurrection would take place.[21]

Describing the change humanity would undergo at the Omega Point, Teilhard said that we would become "some sort of Trans-Human at the ultimate heart of things."[22]

Teilhard wrote this in 1947, several years before Julian Huxley supposedly coined the term "transhuman"—and even Teilhard wasn't the first to use it. Its first appearance was over a century earlier, in one of the first translations of Dante's *Divine Comedy* into English.[23] "After completing his journey through Paradise and ascending into the spheres of heaven, [Dante] describes the process by which his human flesh is transformed," O'Gieblyn writes. "In the end he is forced to make up an entirely new word, *transumanar*, which means roughly 'beyond the human.' When Henry Francis Cary translated the book into English in 1814, he rendered it 'transhuman': 'Words may not tell of that transhuman change.'"[24] While Teilhard may or may not have had Dante in mind when he used the word "transhuman," Huxley definitely had Teilhard in mind when he used it. According to O'Gieblyn, "Teilhard was, not coincidentally, close friends with Julian Huxley, who succeeded in making the priest's ideas mainstream." In the 1951 lecture where Huxley first used the word transhumanism, "Huxley was essentially proposing a nonreligious version of Teilhard's ideas."[25] Transhumanism, then, has always been about ascending to the heavens to live forever, as O'Gieblyn notes. "Most transhumanists are outspoken atheists, eager to maintain the notion that their philosophy is rooted in modern rationalism and not in fact what it is: an outgrowth of Christian eschatology." The dream is always the same, and has been for thousands of years.

* * *

One of the most unfortunate articles to ever appear in the *New York Times* fashion section is a "confirm or deny" interview that Maureen Dowd conducted with Peter Thiel in early 2017, during his stint on then president-elect Trump's transition team. "You like 'Star Trek' more than 'Star Wars,'" she asked. "Deny," he answered. "I like 'Star Wars' way better. I'm a capitalist. 'Star Wars' is the capitalist show. 'Star Trek' is the communist one. There is no money in 'Star Trek' because you just have the transporter machine that can make anything you need.[26] The whole plot of 'Star Wars' starts with Han Solo having this debt that he owes and so the plot in 'Star Wars' is driven by money."[27]

Thiel's reading of *Star Wars* is strange. It's a stretch, at best, to say that the whole plot—even if we're just looking at the first movie—is driven by Han Solo's debt to Jabba the Hutt. Similarly, it's a stretch to call *Star Trek* communist—and it's interesting that Thiel doesn't seem to like the idea of everyone having whatever they need for free.

Despite his apparent difficulty with understanding science fiction, Thiel thinks it's an important source of inspiration. At the 2009 Singularity Summit, Thiel was one of the panelists in a discussion on "Changing the World," alongside Eliezer Yudkowsky and longevity researcher Aubrey de Grey. In response to a question about what things people can build in order to change—and save—the world, Thiel told the audience that "there are many different things that one could be developing. . . . If you wanted to have a menu, I would just give you the list of science fiction books from the '50s and '60s and go through those as starting points. Development of the oceans, development of the deserts, development of outer space, robots, nanotech, biotech, AI."[28] This is another bizarre answer that seems to betray a misunderstanding of science fiction. "Development of the deserts" sounds like it's probably a reference to the 1965 science fiction epic *Dune*, by Frank Herbert. But the message of *Dune* certainly isn't "develop the deserts" any more than *Star Wars* was meant as a guide to building a space empire.

The time period of science fiction Thiel picks out is even more telling. The 1950s and '60s are the middle and end of the Golden Age of science fiction, which started with pulp sci-fi magazines in the very late 1930s like *Astounding Science Fiction*. The authors who dominated this period—such as Isaac Asimov, Arthur C. Clarke, and Robert Heinlein—were almost all white men, and they wrote primarily about a future in space. Asimov's stories were often centered around robots, space empires, or both, with nuclear power depicted as a nigh-limitless energy source used for everything from rockets to radios. Heinlein's stories frequently had a flavor of Ayn Rand in space, usually featuring a self-reliant, polymath male hero dabbling in eugenics or undermining workers on strike for a living wage.

It was Clarke, more than the rest, who dealt in immortality. He was deeply influenced by Tsiolkovsky's cosmist philosophy. Clarke's two most famous novels—*Childhood's End* and *2001*—both feature humans evolving into deathless, spacefaring creatures of pure mind, transcending the need for physical bodies, composed of patterns of energy. (The title of *Childhood's End* is a direct reference to the Tsiolkovsky quote about the Earth being the cradle of humanity. Clarke also referenced Tsiolkovsky by name in his later novel *Rendezvous with Rama*.)[29] In both novels, these evolved humans take their place alongside similarly transcendent aliens of ancient lineage, with technology enabling godlike powers of creation, destruction, and transformation. "Any sufficiently advanced technology is indistinguishable from magic," said Clarke in his most famous dictum. In the meantime, before humanity's technological apotheosis, Clarke saw other reasons for taking to space—reasons echoed by Jeff Bezos and J. Storrs Hall decades later. "Interplanetary travel is now the only form of 'conquest and empire' compatible with civilisation," Clarke wrote in 1951. "Without it, the human mind, compelled to circle for ever in its planetary goldfish-bowl, must eventually stagnate."[30]

Not all of the authors of the Golden Age shared Clarke's optimism about technology and the transcendental possibilities of space. *The Martian Chronicles*, by Ray Bradbury, uses Mars as a setting to explore and critique then standard myths about colonialism and the American frontier. But more trenchant commentary on science fiction tropes largely had to wait until the mid-1960s and the rise of the New Wave, a set of sci-fi authors who wanted to push the genre past its pulp-magazine origins by telling richer stories about a broader range of subjects. These authors included Samuel Delany, Harlan Ellison, Roger Zelazny, Philip K. Dick, and, perhaps most famously, Ursula K. Le Guin. The New Wave authors were interested in widening the horizons of the genre. They wanted to interrogate contemporary notions of progress; they wanted to tell stories about people who weren't straight white men; they wanted to think about politics and class and gender and how they intersect with technology and culture. In time, the New Wave led to the cyberpunk authors of the 1980s and '90s (such as William Gibson, Bruce Sterling, and Neal Stephenson) who were interested in questions about how the wealthy might use technology—especially computer technology—to further concentrate money and power. Small wonder, then, that Thiel's willingness to use science fiction as a blueprint for the future ends around 1969. The first *Star Wars* movie came out almost a decade after that, but that movie was intentionally patterned after the science fiction from thirty years earlier, when George Lucas was a kid. In many ways, *Star Wars* is more like a fantasy movie set in space than a science fiction movie, right down to its setting of long ago and far, far away. But Thiel seems to have missed a crucial detail about *Star Wars* too. Despite Thiel's insistence that *Star Trek* is the communist story, Lucas based the rebel heroes of *Star Wars* on communists. "They were Vietcong," Lucas said.[31] "It was really about the Vietnam War."[32] When asked about the origins of the evil Emperor Palpatine, Lucas gave a simple answer. "He was a politician. Richard M. Nixon was his name."[33]

Thiel's issues with interpreting science fiction reflect a more fundamental problem, one shared by many other tech billionaires: treating science fiction as a forecast, an attempt to predict the future, or depict a desirable one. "Strictly extrapolative works of science fiction generally arrive about where the Club of Rome arrives: somewhere between the gradual extinction of human liberty and the total extinction of terrestrial life," wrote Le Guin in 1976. "Science fiction is not predictive; it is descriptive. . . . Prediction is the business of prophets, clairvoyants, and futurologists. It is not the business of novelists. A novelist's business is lying. . . . I write science fiction, and science fiction isn't about the future. I don't know any more about the future than you do, and very likely less."[34] Nearly half a century later, Charles Stross echoed Le Guin with a more explicit warning. "I—and other SF authors—are terrible guides to the future. Which wouldn't matter, except a whole bunch of billionaires are in the headlines right now because they pay too much attention to people like me."[35] The tech billionaires "read science fiction in their childhood and [they] appear unaware of the ideological underpinnings of their youthful entertainment: elitism, 'scientific' racism, eugenics, fascism and a blithe belief today in technology as the solution to societal problems."[36]

As a professional—and successful—science fiction author, Stross says he's "spent a lot of time lifting up the rocks in the garden of SF to look at what's squirming underneath."[37] One of the slimy grubs he found there was cosmism. Tsiolkovsky didn't just influence Clarke. Tsiolkovsky's quote about Earth as the "cradle" we're destined to outgrow has long been iconic among space enthusiasts and science fiction authors, a shibboleth and a whispered hope. Yet Tsiolkovsky didn't just want humanity to go to the stars. Like Fedorov, he wanted humanity to reorder the cosmos, creating a rationalized universe devoid of anything natural. In this "utopia," tasks would be assigned to different peoples based on the supposedly inherent aptitudes of their race, and the disabled would be killed.[38] Imported across the Atlantic by rocketry enthusiasts and early science fiction authors, Tsiolkovsky's cosmist vision of universal colonialism and eugenics* meshed well with existing ideas in American culture, like Turner's frontier thesis. This set the tone for American science fiction. "There was an implicit ideology attached to this strain of science fiction right from the outset: the American Dream of capitalist success, mashed up with progress through modern technology, and a side-order of frontier colonialism," Stross writes. "There's been a tendency in American SF, ever since those early days, to be willfully blind to the political implications of the shiny toys."[39] Instead, billionaires explicitly use science fiction as a blueprint—Musk even tweeted that "science-fiction should not be fiction forever!"—heedless of the intent or meaning of the works that

* Cosmism, with its promise of eternal life in space through technology—and its colonialist and eugenicist logic—has a clear ideological link to modern movements like transhumanism and singularitarianism. Through cosmism's influence on twentieth-century science fiction, the link is historical as well. Timnit Gebru and Émile Torres have dubbed this set of related ideologies (traced throughout this book) the TESCREAL bundle: Transhumanism, Extropianism, Singularitarianism, Cosmism, Rationalism, Effective Altruism, and Longtermism. Gebru and Torres have done extensive work linking these ideologies to each other and to the core of racist logic they share. While I agree with most of their conclusions (as does Stross), I haven't made use of their terminology in this book for two main reasons: the acronym is filled with jargon that takes some work to unpack and is itself not particularly legible; and by the time I encountered their work on TESCREAL, I was already deep into the writing of this book. For an introduction to their work on this subject, see Gebru and Torres, "The TESCREAL Bundle: Eugenics and the Promise of Utopia Through Artificial General Intelligence," https://dx.doi.org/10.5210/fm.v29i4.13636.

they're blindly emulating.[40] "Did you ever wonder why the 21st century feels like we're living in a bad cyberpunk novel from the 1980s?" Stross asks. "It's because these guys [tech billionaires] read those cyberpunk novels and mistook a dystopia for a road map."[41]

But it's not much of a surprise that Thiel, Bezos, Musk, and their ilk seem to have trouble understanding science fiction. The culture of the tech industry—especially start-up culture—doesn't value the sort of skills that someone might learn in an introductory college English course, a history class, or really anything at all in the humanities. The old line about a liberal arts education is that it teaches you to think. Thiel has made it clear he doesn't believe this. He encourages potential tech founders to avoid college by giving $100,000 to about two dozen "Thiel fellows" each year, contingent on the recipient skipping or dropping out of college. "College can be good for learning about what's been done before, but it can also discourage you from doing something new," the Thiel Fellowship website claims. "Each of our fellows charts a unique course; together they have proven that young people can succeed by thinking for themselves instead of following a traditional track."[42] (One wonders whether the Thiel fellows' success has less to do with independent thought and more to do with their access to a network of Silicon Valley power players, provided as a part of the fellowship.) Thiel isn't the only one to see college as a waste of time for tech founders. Going back at least as far as Steve Jobs and Bill Gates, there's an idea halfway between a tradition and a mythology in the tech industry that successful start-up founders must gain entry to elite colleges only to drop out after a year or two. Those that do stay usually major in a STEM field (science, technology, engineering, and mathematics). This homogeneous, STEM-focused intellectual background in tech start-up culture, and in Silicon Valley more generally, engenders *humanities denial*, a systematic and sometimes willful ignorance of the arts and humanities.

Nowhere is humanities denial more apparent in the tech industry than in its attitude toward history. "I don't even know why we study history," says Anthony Levandowski, cofounder of Google's self-driving car division, now known as Waymo. "It's entertaining, I guess—the dinosaurs and the Neanderthals and the Industrial Revolution, and stuff like that. But what already happened doesn't really matter. You don't need to know that history to build on what they made. In technology, all that matters is tomorrow."[43] Levandowski may not be the most reliable narrator: In 2020, he pleaded guilty to stealing trade secrets from Waymo (Trump pardoned him the next year).[44] He also founded a religious organization dedicated to "the realization, acceptance, and worship of a Godhead based on Artificial Intelligence."[45] But the sentiment he expressed is only a slightly more extreme version of a fairly common view within the tech industry. Silicon Valley is "a place that likes to pretend its ideas don't have any history," writes Adrian Daub, Stanford professor and author of *What Tech Calls Thinking*.[46] Thinking about history would force the leaders of the tech industry to confront uncomfortable questions about their own culture and the unspoken assumptions at work behind the grandiose statements they make.

Taking history seriously would also force tech billionaires to reexamine the role models and goals they choose for themselves and how they portray their own works. In a promotional video for Blue Origin, Bezos addresses an unseen audience, repeating his same spiel about the need to go into space to allow a future with constant growth in energy used per capita, set against dramatic footage of his company's rockets in flight. As the video ends with a shot of a parachuted capsule setting down in desert scrub, Bezos intones: "Anything you set your mind to you can do. Von Braun said after the lunar landing, 'I have learned to use the word 'impossible' with great caution,' and I hope you guys take that attitude about your lives."[47] Bezos also trotted out the same quote in a podcast interview several years later, calling it "one of the great" quotes from the rocketry pioneer.[48] But Bezos is either unaware or unconcerned with Wernher von Braun's history. Before his time at NASA, von Braun was building V-2 rockets for the Nazis, forcing Jews and others in concentration camps to assemble the flying bombs for Hitler's attacks on the United Kingdom and western Europe as the Allies closed in. Von Braun's rockets killed thousands during the late stages of World War II, but even more prisoners died building the rockets in the first place.[49] None of this is a secret; it wasn't even a secret during the *Apollo* missions. It's possible that Bezos knows about this history and doesn't care. But it's also quite possible that he just doesn't know. Humanities denial makes it easy to ignore inconvenient facts, or remain blissfully unaware of them in the first place.

Humanities denial is enabled—and enables—another affliction that's endemic within the tech industry: "engineer's disease," the belief that expertise in one field (usually in STEM) makes you an expert on everything else too.[50] Or, put another way, there's only one thing that's actually difficult, and you already know it, so everything else must be easy. Software developer Maciej Cegłowski explains the source of this hubris:

> As computer programmers, our formative intellectual experience is working with deterministic systems that have been designed by other human beings. These can be very complex, but the complexity is not the kind we find in the natural world. It is ultimately always tractable. Find the right abstractions, and the puzzle box opens before you. . . . But as anyone who's worked with tech people knows, this intellectual background can also lead to arrogance. People who excel at software design become convinced that they have a unique ability to understand any kind of system at all, from first principles, without prior training, thanks to their superior powers of analysis.

> Success in the artificially constructed world of software design promotes a dangerous confidence.[51]

The logical end point of this line of thought is the simulation hypothesis, the idea that we all live inside a computer already. Elon Musk is probably the most famous proponent of this idea. "We're clearly on a trajectory to have games that are indistinguishable from reality, and those games could be played on any set top box or on a PC or whatever, and there would probably be billions of such computers or set top boxes, it would seem to follow that the odds we're in base reality is one in billions," Musk claims. "Tell me what's wrong with that argument. Is there a flaw in that argument?"[52]

There are quite a few flaws. Musk isn't referring here to something like *The Matrix*, where our bodies have been hooked up to a computer simulation. He's talking about something more radical: the idea that we ourselves, and everything in our universe, are actually the products and inhabitants of a computer simulation, existing only in the memory of a computer in an entirely different reality, run by some other entity, human or otherwise. Musk's argument echoes one made by Nick Bostrom in a 2003 paper titled "Are You Living in a Computer Simulation?" There, Bostrom concludes that, given the computing power available to a post-Singularity civilization, it's likely that such future humans (or transhumans) would have the ability to run myriad "ancestor simulations," simulating earlier phases in their society's history. Life inside an ancestor simulation would look much like life in a pre-Singularity society such as ours, so it's not immediately obvious whether we live in one of those many simulations or in the real world. And because there could be so many of those simulations, Bostrom concludes that if humanity is likely to reach the Singularity, we're much more likely to be the inhabitants of a simulation rather than the singular real world.

Even if we accept Bostrom's contention that the brain can be simulated on a computer—and also accept Bostrom's estimate of the computing power needed to do so—the simulation argument still has to contend with the problem that nobody knows how to build a computer that could run an ancestor simulation or one of Musk's hyperrealistic video games. With the end of Moore's law at hand, it may never be possible to build such a computer. Even if one could eventually be built, it's quite unreasonable to claim that such computers will inevitably become so ubiquitous—and so frequently used to perform ancestor simulations or play phenomenally detailed video games—that we ourselves are overwhelmingly likely to be the inhabitants of such a simulation.

Despite these counterarguments, the simulation hypothesis is widely entertained within the tech industry, as well as the rationalist and EA communities. It's not hard to see why. Instead of the total destruction of nature offered by the Singularity and the maximalist futures of the longtermists, the simulation hypothesis goes one step further. It entails the *denial* of nature: what appears to be the natural world is in fact simply a computer program. It was authored by humans—or, more properly, entities like humans but vastly more powerful—and it was made for us. We can do with it as we like. (It is hard not to read into this shades of Genesis: We are made in the creator's image, the world was made with us in mind.) If we live in a computer simulation, then expertise in software engineering really is expertise in everything. The simulation hypothesis reveals the natural complexity of the world as an illusion, with the artificial complexity that Cegłowski spoke of as a more fundamental—and tractable—substrate.

This is not a new idea. The eighteenth-century Irish bishop and philosopher George Berkeley wrote of reality as existing solely in the mind of the Christian God, a realm of pure rational thought. There are more recent antecedents as well. "Tlön may well be a labyrinth, but it is a labyrinth forged by men, a labyrinth destined to be deciphered by men," wrote Jorge Luis Borges of the artificial yet intricate world of Tlön in his short story "Tlön, Uqbar, Orbis Tertius." "Spellbound by Tlön's rigor, humanity has forgotten, and continues to forget, that it is the rigor of chess masters, not of angels." In the story, Tlön is devised by a ring of intellectuals who seek to replace the real, natural world with an artificial one, an "ordered planet" where ideas have primacy over material things and history can be changed at will—like a computer program.[53] Borges wrote the story during the early stages of World War II, and at the close of the story he explicitly compares Tlön with the orderly paranoia of Nazism and anti-Semitism. Small wonder, then, that when Curtis Yarvin started a company with Peter Thiel's backing, he chose Tlön for its name.

The comforting lie behind both Tlön and the simulation hypothesis is that they place us at the center of creation, imbued with cosmic purpose, alive at the fulcrum of universal history. "This is deeply embarrassing to admit but one reason I always found the simulation hypothesis strikingly plausible is that it would explain how I just happened to find myself hanging around who I considered to be plausibly the most important people in human history," wrote Qiaochu Yuan, the former rationalist. "Because you'd think if our far future descendants were running ancestor simulations then they'd be especially likely to be running simulations of pivotal historical moments they wanted to learn more about, right? And what could be more pivotal than the birth of AI safety?"[54]

But Cegłowski thinks that the simulation hypothesis reflects a deep fear, rather than a deep yearning. "Fantasies of control come with a dark side," he says. "For a computer programmer, [living in a simulation is] the ultimate loss of control. Instead of writing the software, you are the software."[55] There has been at least one attempt to reassert that control: in 2016, the *New Yorker* reported that there were "two tech billionaires [who] have gone so far as to secretly engage scientists to work on breaking us out of the simulation."[56]

Fundamentally, the simulation hypothesis rests on a bevy of unexamined assumptions about the future of humanity. Mostly, those assumptions are about the durability of ideas held by people here and now (sometimes not even by that many people), ideas about everything from economics to sociology to cosmology, most of which are not informed by actual expertise in any of those fields. But thinking critically about these ideas isn't really the point. If it was, all the serious issues with the simulation hypothesis would have ended the conversation about it long ago. The idea is appealing because it implies that tech is truly all that matters, that reality runs on the legible logic of computer programs rather than the obscure mechanics of a world we never made. It also carries the promise of transcending reality itself by escaping the simulation, though how that would work when we ourselves are purportedly creatures of the simulation isn't clear. But that objection doesn't really matter either. It's just vibes all the way down.

* * *

In the spring of 2023, I was invited to a small dinner hosted by the founder and CEO of an AI-based tech start-up. (The names of the people and companies involved in this story have been withheld to protect their anonymity.) The dinner was part of an event celebrating the company's launch of a new AI product, as well as the fact that a major VC firm had just given the company a large sum of money. I had no direct connection to the start-up, the VC firm, or the founder. I was, frankly, surprised to be there. The dinner was held at a small restaurant, widely regarded as one of the finest in the city. There were roughly a dozen people in attendance, including an executive from a well-known tech company, the leader of a major product team at another AI company, an investor in the start-up famous for his work in an unrelated field, and a few people working in senior positions at the biggest tech companies. Everyone was white; there was only one woman in the room. We were seated around a table in a private dining room. I sat to the right of the start-up founder and to the left of one of the only other people there who had never worked in tech, a philosopher specializing in consciousness. The philosopher and I chatted for a bit while the rest of the guests got settled, and then the founder spoke. He announced that we'd be having one large collective conversation for most of the evening, and that it would revolve around—what else?—AI. Once we finished introductions, we started with a single question: When would the first AGI be developed? The founder asked us each to go around the table and make a specific prediction, involving a well-defined period of time. To my relief, he started with the person sitting on his left.

The conversation soon got bogged down in the question of what AGI is, until someone—I don't recall who—suggested a working definition: a machine that can reproduce any economically productive activity done by a human. (I later found out that there's an almost identical definition of AGI in the OpenAI charter.)[57] I thought that was a pretty bad definition. It struck me as rounded down, both too vague and too narrow. What's an economically productive activity? Why focus on economic productivity in the first place? What about human activities that aren't economically productive at all, like daydreaming or going for a long walk with a friend? Besides, there's no single person who can perform every economically productive activity humans engage in, for any reasonable definition of "economically productive."

But I held my tongue for the moment, because it wasn't my turn yet—and because I quickly realized that I had a different problem to deal with. By the time we were halfway around the table, nobody had given an answer larger than ten years. The most common answer was that the development of the first AGI was around five years away; a couple of people went even lower than that. One person said that the number of parameters in GPT-4 was only one hundred times smaller than the number of neurons in the human brain, and that it shouldn't take more than ten years to get that last factor of one hundred, so his guess was ten years. These people knew a lot about these AI systems. At least two people at the table had led teams building AI products, and I knew that one of those people had a strong academic background in machine learning. I don't—I'm a physicist by training. I've never worked in tech. Nearly everyone else at the table did or had done so. One of the few who didn't—the investor, sitting several places down the table to my right, almost opposite the founder—guessed five years, saying that he didn't know nearly as much about AI as the other people at the table, and so he was just going to follow their lead in making a guess. It wasn't the worst idea. What would I say? By then, I'd already written part of a draft of this book, and I went over my reasoning again in my head. It sounded more hollow than it had on the page. What if they were right, and I was wrong?

If this had been a scene in a movie or a chapter in a novel, then this would have played out differently. I would have taken a defiant stand against the consensus of the room, there would be shouting and possibly crying, and then I would emerge victorious. Instead, I didn't get to find out if I had the courage to stand alone against the rest of the room. The philosopher relieved me of that. He explained that he had no idea how long it would take to build an AGI that wasn't conscious, but that if AGI requires consciousness, then he thought we'd have to reinvent science from the ground up in order to successfully build one. That, he said, would take a long time. Pressed for a number, he guessed five thousand years.

There was a stunned silence as the assembled group chewed on the philosopher's words. Then it was my turn. I explained that I didn't think the definition we had of an AGI was very good, but that if it was going to be able to do everything a human can do, it would probably have to be conscious. The good news, I said, was that I was more optimistic than the philosopher about how long that would take, because while I did think that we'd have to use fundamentally different machines, I didn't think we'd have to reinvent all of science to do it. So it might only take three or four centuries—a bargain, really.

I was shocked by how little pushback I got from the rest of the group for my guess. Mostly, they just nodded and moved on. (The most substantial feedback came from the one woman at the table, who was sitting on the other side of the philosopher. She told us she wished she'd given a bigger answer, too, but that she hadn't felt comfortable doing so in the face of the unanimously optimistic group when her turn had come.) I'd reviewed my arguments when I was trying to steel myself, before the philosopher jumped in, because I'd thought I would need them to justify my position. But when I thought over what I'd heard others say, the most substantial argument given was the one about the number of parameters in GPT-4 as compared to the number of neurons in the human brain, and that argument had made no mention of the fact that neurons are not much like the internal parameters of a large language model. Aside from that, mostly I'd just heard people say stuff like "Things have been moving quickly recently, and I expect that to continue," or "I agree with everyone else."

I got to thinking about what it would be like to work with these people, to be surrounded by them and people like them for most of the day on most days. They were nice, and the conversation was pleasant. I liked them. But I also remembered that feeling I had before the philosopher spoke, the sense that I might have quailed in the face of the majority opinion. I don't think I would have, though of course I can't be sure. But if I worked with these people, that might have been different. If, instead of just a dinner, I spent forty hours a week or more with them, if they became my friends outside of work, would I still feel comfortable voicing my dissenting opinion? If they remained unanimous and confident in their position, and I full of doubt about mine (as I tend to be), would I eventually decide they might be onto something? I can't know the answer to that either. But it did leave me wondering: With so much confusion at the start of the conversation about the definition of AGI itself, and given the shaky definition they eventually settled on, how could they have been so unanimously sure that, whatever it was, it was coming soon?

* * *

Groupthink goes a long way toward explaining the popularity of ideas like the Singularity, AGI, and space colonization despite the cornucopia of arguments against them.[58] But it's surprising, on the face of it, to see so much groupthink in the tech industry, with its culture that claims to prize independent thought and contrarianism. Andreessen's manifesto decried the very idea of expertise. Thiel has repeatedly stressed the importance of being contrarian in his own life, and the vaunted power of independent thought is the supposed strength of his Thiel fellows. The lone founder finding success and wealth by swimming against the current is a powerful image in start-up culture.

But contrarian thinking can itself turn into a kind of groupthink when it becomes reflexive, automatically dismissing nearly anything perceived to be widespread conventional wisdom or expert opinion, no matter its source. This attitude—which economist Adam Ozimek calls a "brain-rotting drug"—is widespread among the leaders of Silicon Valley. "Tech bros appear to be especially susceptible to brain-rotting contrarianism," writes Paul Krugman. "Their financial success all too often convinces them that they're uniquely brilliant, able to instantly master any subject, without any need to consult people who've actually worked hard to understand the issues." This is engineer's disease, though Krugman doesn't name it as such. "In many cases they became wealthy by defying conventional wisdom, which predisposes them to believe that such defiance is justified across the board."[59] The problem with such reflexive contrarians, as Ozimek points out, is that "they lose the ability to judge others they consider contrarian, become unable to tell good evidence from bad, a total unanchoring of belief that leads them to cling to low quality contrarian fads. As soon as 'experts are wrong' becomes their guidestar . . . their ability to gauge reality becomes extremely restricted. It's like mental glaucoma."[60]

This kind of reflexive contrarianism is tightly linked to the myth of the tech billionaire as genius. Silicon Valley is an enormous casino. For every start-up that becomes a Facebook or a Google—a "unicorn" in tech slang—there are hundreds that crash, sputter, or simply vanish.[61] It may be the case that some successful founders and venture capitalists have a true knack for spotting start-up ideas that are likely to do well at a given time. But it's surely the case that many—perhaps most—tech billionaires were simply at the right slot machine at the right time through pure happenstance. Winning can make you drunk on confirmation bias: your good fortune is proof that everything you did along the way was the right move. "Success all too easily feeds the belief that you're smarter than anyone else," writes Krugman.[62] Or, as the tech entrepreneur and writer Anil Dash puts it, "I must be smart, look how rich I am."[63]

Yet by and large, the tech elite don't see the perils of this way of thought. Rather than recognize that their beliefs might be the result of the kinds of "low quality fads" that Ozimek wrote about, they look for more confirmation—and if they have enough money, they can simply fund their own echo chamber to provide them with the justification they seek for their contrarian views. "It's impossible to overstate the degree to which many big tech CEOs and venture capitalists are being radicalized by living within their own cultural and social bubble," writes Dash. And, he warns, that echo chamber combines with knee-jerk contrarianism to reinforce the impression that they are at odds with the rest of the world. "Their level of paranoia and contrived self-victimization is off the charts, and is getting worse now that they increasingly only consume media that they have funded, created by their own acolytes. In a way, it's sort of like a 'VC Qanon.'"[64]

This is how the ideas central to the ideology of technological salvation survive despite the specious arguments in their favor and the mountain of evidence against them. Reflexive contrarian thought closes off your mind, making it impossible to change your opinions. It insulates you from learning anything new, keeping you at the center of an epistemic cell of your own making, secure in the knowledge that you know best, whatever the rest of the world may say.

* * *

One of the reasons it's easy to dismiss *Star Trek* as facile or childish is that it's not particularly interested in being subtle. (This is especially true of the original series and *The Next Generation.*) In the episode "The Measure of a Man," Captain Picard (Patrick Stewart) is talking with Guinan (Whoopi Goldberg) about a court case he's enmeshed in regarding the rights and status of Data, a member of the crew and his friend, who is an android. Guinan points out that if the judge rules that Data is merely property, that will lead to generations of "disposable" androids, forced to do the work nobody else wants to do. A Black woman is saying this to a white man; the subtext—which has been rising higher and higher all episode and is barely below the surface at this point—is clear. But *Star Trek* is rarely satisfied with letting subtext remain subtext, no matter how close to the surface it lies. Upon hearing Guinan talk about disposable workers, Picard looks almost directly into the camera and says, "You're talking about slavery."

This level of allegorical bluntness is the norm for the franchise. In the movie *Star Trek VI,* a thinly veiled metaphor for the end of the Cold War, Mr. Spock tells Captain Kirk that "only Nixon could go to China." In the episode "Patterns of Force," Kirk and Spock—played by

William Shatner and Leonard Nimoy, both Jewish actors—overthrow the government of a planet ruled by literal space Nazis, complete with swastikas. And even that wasn't direct enough: in the episode "The City on the Edge of Forever," widely regarded as the best episode of the original series, Kirk and Spock have to follow a temporarily deranged Dr. McCoy back in time to keep him from accidentally letting the Nazis win World War II.

Yet *Star Trek*'s lack of subtlety can be a virtue at times. Direct morality plays have their place. (And given the existence of *Star Trek* fans who are vocally upset with the "wokeness" of new *Trek* shows like *Discovery* and *Strange New Worlds*, it seems that some viewers still managed to miss the point.) Gene Roddenberry, the creator of *Star Trek*, wanted a diverse crew on the bridge of the *Enterprise* specifically to depict a future without systemic racism and sexism. This has been dismissed as tokenism by some critics, and reasonably so, especially since the captains and first officers on the first two *Star Trek* shows were all white men. (To Roddenberry's credit, he wanted a woman to be the first officer on the original show; he was overruled by the network executives. Starting with *Deep Space Nine*, later entries in the franchise portrayed a wider range of roles for women and minorities, including command.)[65]

But representation still matters, and the diverse casting of *Star Trek*—especially in the original series, airing in the second half of the 1960s—served as an inspiration to a great number of people who weren't white men. After the end of the first season of the original show, Nichelle Nichols, the Black woman playing Lieutenant Uhura, had decided to quit. But she changed her mind because a fan of the show, Martin Luther King Jr., urged her to stay.[66] Whoopi Goldberg later said that watching Nichols on *Star Trek* when she was a child was the first time she'd seen a depiction of the future with anyone Black in it. Mae Jemison, the first Black woman in space, said that *Star Trek* inspired her to become an astronaut. (During her time as an astronaut, Jemison even had a cameo on *The Next Generation.*) *Star Trek* tries—in an imperfect way—to depict a better future, and uses that depiction to hold up a mirror to the present and the past.[67] The core hope of the show is not that we'll find our way to the stars, at least not necessarily; it's that we'll find a way of building a kinder and more inclusive future. The stars are just the setting. They're incidental. The people are what matter.

On some level, I understood that when I was a kid—that's why I connected with the show. I liked the people; I wanted them to be my friends. But I confused the hope for a better world with the setting of the show. I thought that we'd become better people in the future *because* we'd go to the stars, that going to space would somehow make humans and humanity kinder and more noble. I wasn't alone in this. The "overview effect" is the term for the supposedly durable psychological impact of going to space and seeing the Earth as it is, a single fragile planet. "You develop an instant global consciousness, a people orientation, an intense dissatisfaction with the state of the world and a compulsion to do something about it," said Edgar Mitchell, an astronaut who landed on the Moon on *Apollo 14*. "From out there on the moon, international politics look so petty. You want to grab a politician by the scruff of the neck and drag him a quarter of a million miles out and say, 'Look at that, you son of a bitch.'"[68] It's a compelling idea. In 1966, the same impulse prompted Stewart Brand to sell pins saying, "Why haven't we seen a photograph of the whole earth yet?"[69] Two years later, Brand put one of the first such images on the cover of the first *Whole Earth Catalog*, which espoused a peculiarly Californian fusion of the countercultural philosophy of the hippies and a techno-utopian vision of the future. Brand had hoped to shift public consciousness by drawing attention to the image of the whole Earth, and it likely did have some contribution to the nascent environmental movement. (Brand himself would go on to become a fixture of the tech scene in San Francisco for decades; as of 2024, he still is.) But the evidence for the overview effect is thin at best. Mitchell thought that showing Earth from space to politicians would cure the world's ills, but there are astronauts who have become politicians, and none of them found ways to transcend the pitfalls of political life. John Glenn, the first American to orbit the Earth, became a Democratic senator from Ohio; he was implicated in the Keating Five scandal. Harrison Schmitt, the twelfth person to set foot on the Moon, became a Republican senator from New Mexico; he has taken money from fossil fuel companies and denies the scientific fact that humans are causing global warming. (There have been other astronauts turned senators, like Bill Nelson and Mark Kelly. Their political positions are fairly standard for their parties, with no clear evidence of the overview effect at play.) But I didn't know about any of this as a kid. I didn't care about politics. It just seemed like more messy details, weird adult stuff that clearly wasn't as important as space and sending humans there. When I first read Mitchell's quote about the overview effect, it made perfect sense to me—it was further confirmation of what I already wanted to believe. Going to space would fix everything. Technology—specifically, technology to get off the Earth—was the answer to the world's ills.

But then I grew up. Sagan (and others) had taught me that a core precept of science is that nobody is infallible, not even Sagan. By the time I watched *Cosmos*, it was over a decade old; *Star Trek* and the science fiction books I'd been reading were two or three times older than that. Humans had found out much more about space than Clarke, Roddenberry, or even Sagan had known at the time they created their works. And what we'd discovered was mostly bad news for human space exploration. In 1951, Clarke didn't know about the wildly infernal temperatures on the surface of Venus (the explanation for which was uncovered several years later by a PhD student at the University of Chicago named Carl Sagan).[70] In 1966, Roddenberry didn't know that long-term exposure to zero-g environments led to bone-density and muscle-mass loss (though he did give the *Enterprise* artificial gravity). In 1980, Sagan didn't know that the surface of Mars is covered in toxic perchlorate compounds. In fact, nobody knew that last fact when I was watching *Cosmos* in the midnineties—it wasn't discovered until the *Phoenix* Mars lander performed a chemical analysis of Martian surface dust in 2008, and I didn't hear about it until several years after that.[71]

By that time, my love for space had seen me through a PhD in cosmology and into a career as a science journalist, specializing in astronomy and physics. As I'd learned more, and as my interests had broadened, I'd slowly given up on more and more of the idea of a human future in space. But it had happened in the background; I hadn't really been aware of it. Strangely, it was learning of the discovery of perchlorates on Mars—hardly the biggest barrier to large numbers of humans living off of Earth—that finally brought the whole picture into focus for me. *That can't be right*, I thought when I heard the news. *If Martian dust is filled with deadly poisons, how can we live there?* Then all of my training as a scientist, from Sagan to graduate school, sounded an alarm: I was looking at what I wanted the evidence to be, not what it actually was. The perchlorates were there, on Mars, right alongside all the other problems with the Martian environment. Nobody was going to live there, at least not for long. And thinking about it more, I realized that I'd been working my way to that conclusion for years. I'd known that the overview effect was flimsy; I'd known that getting off the surface of the Earth was dangerous, difficult, and expensive; I'd known that there wasn't much of a political, social, or scientific case for sending large numbers of people into space. We evolved here. We're suited for this place. Earth isn't just our cradle—it's our home. As Sagan wrote in 1994:

> On [Earth] everyone you love, everyone you know, everyone you ever heard of, every human being who ever was, lived out their

> lives. The aggregate of our joy and suffering, thousands of confident religions, ideologies, and economic doctrines, every hunter and forager, every hero and coward, every creator and destroyer of civilization, every king and peasant, every young couple in love, every mother and father, hopeful child, inventor and explorer, every teacher of morals, every corrupt politician, every "superstar," every "supreme leader," every saint and sinner in the history of our species lived there—on a mote of dust suspended in a sunbeam. The Earth is a very small stage in a vast cosmic arena. . . . Our posturings, our imagined self-importance, the delusion that we have some privileged position in the Universe, are challenged by this point of pale light. Our planet is a lonely speck in the great enveloping cosmic dark. In our obscurity, in all this vastness, there is no hint that help will come from elsewhere to save us from ourselves. The Earth is the only world known so far to harbor life. There is nowhere else, at least in the near future, to which our species could migrate. Visit, yes. Settle, not yet. Like it or not, for the moment the Earth is where we make our stand. . . . There is perhaps no better demonstration of the folly of human conceits than this distant image of our tiny world. To me, it underscores our responsibility to deal more kindly with one another, and to preserve and cherish the pale blue dot, the only home we've ever known.[72]

Sagan wrote those words when going for a different kind of overview effect with a different kind of picture of the whole Earth: he had urged the *Voyager 1* team to turn the probe around at the edge of the solar system in 1990, just before its cameras powered down, for a final picture of the Sun and planets. Earth appeared as a fraction of a pixel, a dot in a shaft of sunlight. And when Lex Fridman asked Elon Musk to read Sagan's words aloud on his podcast in 2019, Musk paused at the spot where Sagan said that "there is nowhere else, at least in the near future, to which our species could migrate." "This is not true," Musk protested. "This is false—Mars."[73]

I don't say this often, but I have sympathy for Elon Musk, at least in that moment. He was facing the same thing I once was: evidence running counter to a deeply held belief. It's hard in those moments to let the new evidence in, to face the fear that comes with the possibility of being wrong. This is, in principle, what the Center for Applied Rationality tries to help people do, though in practice it seems to spend more effort pushing the rationalist ideology. But I have sympathy for the rationalists too, and for the effective altruists, the longtermists, Kurzweil, Yudkowsky, MacAskill, and the rest of the characters that populate this book, despite how it may appear. I have a lot in common with them: I'm a white guy with a scientific background, I love space and technology, and I'm a huge science fiction nerd. But I can't follow where they want to lead, because I know where that road goes. They're blinkered. Part of that is unearned confidence in the power of technology to cure all ills, and in their own ability to discern the truth regardless of what experts may tell them. But underneath it all, I suspect, is a deep faith in space as the inevitable location of the future, the place we go to become better versions of ourselves. But that's an outdated vision of the future, yesterday's enterprise based on yesterday's ambitions and knowledge. And as much as I have sympathy for these people, I can't lose sight of the sympathy I also have for the people their visions harm. *Star Trek* and Carl Sagan taught me that we need to build a tomorrow for everyone on this mote of dust in a sunbeam, and take care of each other in this fragile place. We aren't leaving Earth. But we already live among the stars.

* * *

There is a final question to consider, one that I've been dreading: If not an immortal future in space, then what?

I don't know. The futures of technological salvation are sterile impossibilities, and they would be brutally destructive if they did come to pass. The cosmos is more than a giant well of resources, and humans are more than siphons sucking it dry. But I can't offer a specific future as an alternative. What I can tell you is that anyone who claims to know the one inevitable future, or the one good path for humanity, is someone who deserves your deepest skepticism. I don't have a comprehensive vision of tomorrow because I don't believe anything like that is genuinely possible. This is especially true when it comes to purported utopias. Human happiness is a complex thing, even for one person; envisioning a permanently, perfectly happy society is generally beyond our capacities. "Nearly all creators of Utopia have resembled the man who has toothache, and therefore thinks happiness consists in not having toothache. They wanted to produce a perfect society by an endless continuation of something that had only been valuable because it was temporary," George Orwell wrote. "The wiser course would be to say that there are certain lines along which humanity must move, the grand strategy is mapped out, but detailed prophecy is not our business. Whoever tries to imagine perfection simply reveals his own emptiness."[74]

Most of the greatest problems facing humanity right now—global warming, massive inequality, the lurking potential for nuclear war—are not driven by resource scarcity or a lack of technology. They're social problems, requiring social solutions. Increased energy usage, increased technological prowess, or even an increase in the amount of "intelligence" brought to bear on these problems (whatever that might mean) isn't likely to solve them. These are political problems, problems of persuasion and justice and fairness. Negotiating durable ceasefire agreements isn't something that AI is likely to help with; sending vast numbers of humans to space won't end violence in the Middle East; doubling humanity's energy consumption isn't going to fix political polarization in the United States or stem the rise of fascism worldwide. If we want a future that puts people first, we need to recognize that there are no panaceas, and likely no utopias either. Nothing is coming to save us. There's no genie inside a computer that will grant us three wishes. Technology can't heal the world. We have to do that ourselves.

None of this is to say that technology is useless at solving problems—just that it must be directed, that we must make choices about what we want technology to do as part of the solution to some of our problems, rather than presuming the right technology will come along to solve all of them entirely. Like all human activities, developing technology is full of contingency and choice. Weirdly, Peter Thiel understands this better than most tech billionaires. "The future of technology is not pre-determined, and we must resist the temptation of technological utopianism—the notion that technology has a momentum or will of its own, that it will guarantee a more free future, and therefore that we can ignore the terrible arc of the political in our world."[75] But Thiel's ideas of "a more free future" and what constitutes "the terrible arc of the political" are wildly different from my own. Like his fellow billionaires, Thiel has a habit of ignoring or doubting scientific facts that run counter to his worldview. (He even funded an online magazine that promoted creationism.)[76] Thiel's idea of "freedom" seems to consist of free markets and not much else, so perhaps it's no surprise that the future of such a worldview is profoundly inhumane. Thiel, like many other people in this book, is bound to an ideology that blinds him to the world around him. He is, in a sense, too idealistic—too committed to "the faith of [his] teenage years," as he puts it, that death is avoidable and markets should be free of any regulation.[77] Neither of those ideas can survive contact with the real world, where people die and markets are merely a limited tool invented by humans, not a fundamental feature of the universe.

Such criticism is easy, though. While I do genuinely believe that creating a complete vision of a good future isn't something people do well—it's certainly not something I do well—it's also a convenient position for me to hold while I'm taking potshots at the leaders of Silicon Valley and their kept futurists. So here's a specific policy proposal, one that's even endorsed by a major science fiction author. "There should be no such thing as billionaires," Kim Stanley Robinson tells me. "The Midas touch is not a happy thing—if you touch people and they turn to gold, then this is a serious barrier to intimacy. So it would be doing them a favor to tax them out of existence. The Republican Congress of 1953 under Dwight Eisenhower was pretty good at this—incomes over $300,000 a year were taxed at 92 percent for the overage. This was a society that understood the rich to be parasites and fools. We live in a stupider time, but we could change that."[78] The fact that our society allows the existence of billionaires is the fundamental problem at the core of this book. They're the reason this is a polemic rather than a quirky tour of wacky ideas. Without billionaires, fringe philosophies like rationalism and effective accelerationism would stay on the fringe, rather than being pulled into the mainstream through the reality-warping power of concentrated wealth. As Robinson says, there's no reason we as a society have to put up with the continued existence of billionaires. Consider a wealth tax where the top bracket takes 100 percent of personal net worth above, say, $500 million. $500 million is an enormous amount of money; you could spend $3,000 a day for a hundred years and still have nearly $400 million left over. With $500 million in the bank at a measly 1 percent interest rate, you'd still have an income of $5 million a year without ever diminishing the bulk of your wealth. There is simply no real need for anyone to have more money than half a billion dollars—and there's ample justification for returning all personal wealth over that amount back to society.

Nobody earns their wealth alone. A functioning and stable market is something that can only exist within a society that has working infrastructure, health care, education, and everything else it takes to make a modern thriving country. The tech industry itself was built on the back of massive public spending by the United States: Government investment in basic science created the academic research environment that was crucial for the development of Silicon Valley in the first place, in the shadow of Stanford University. The government, in the form of NASA, provided most of the demand at the first semiconductor companies during the space race. Even the internet itself was first created as a government project, known as ARPANET.[79] Unlike private investors, the US government hasn't seen a direct return on that investment. If you must, you can think of such a wealth tax as the government recouping its investments in roads, trains, public schools, basic scientific research, and all the other things that have made such massive wealth possible in the first place. That tax would also be an investment in the future of technology. It would enable far more fundamental scientific research, fighting against the diminishing returns of the past decades to keep finding innovative solutions to technical problems.

Eliminating billionaires would also be an investment in the political stability that makes prosperity possible. Over eighty years ago, Louis Brandeis warned that "we may have democracy, or we may have wealth concentrated in the hands of a few, but we can't have both."[80] The past decade has provided ample proof that Brandeis was right: concentrated wealth has eroded democracy in the United States and around the world. Peter Thiel has aided that erosion with his support of Trump and other far-right politicians. The antidemocratic ambitions of tech billionaires extend through Sam Altman's power fantasy of his own ascension to king of the world straight to the permanent galactic fascism of Marc Andreessen. While those specific visions are implausible, the fact that billionaires harbor such dreams is an indication of the risks that we as a society are taking by allowing billionaires to happen. They will keep looking for ways to extend their control over the world unless they are curtailed.

Their dreams are dreams of endless capitalism of the most brutal sort, because they know that such a system would allow them to win still more money and power. This is another reason it's difficult to imagine a future other than the ones they promote: as the saying goes, it's easier to imagine the end of the world than the end of capitalism. But eventually—and probably sooner rather than later—the present growth-obsessed form of capitalism must come to an end. And it can end without ending the world, as Ursula K. Le Guin reminded us. "We live in capitalism. Its power seems inescapable. So did the divine right of kings. Any human power can be resisted and changed by human beings."[81]

This is why the tech billionaires tell us their futures are inevitable: to keep us from remembering that no human vision of tomorrow is truly unstoppable. They want to establish a permanent plutocracy, a tyranny of the lucky, through their machines. They are too credulous and short-sighted to see the flaws in their own plans, but they will keep trying to use the promise of their impossible futures to expand their power here and now. "I think hard times are coming when we will be wanting the voices of writers who can see alternatives to how we live now and can see through our fear-stricken society and its obsessive technologies to other ways of being, and even imagine some real grounds for hope," Le Guin said. "We will need writers who can remember freedom, poets, visionaries—the realists of a larger reality."[82] This act of imagination Le Guin suggested is vital. Steeped in the political realities of today's world, it feels impossible to imagine actually accomplishing a change even as modest as taxing billionaires away. It is that freedom, the freedom to imagine alternative ways the world could be, that the tech billionaires' rhetoric of inevitability tries to wrest away from us. We must remember that, in truth, their visions aren't inevitable—they're all but impossible. There are other tomorrows, lush and desolate, gorgeous and harrowing, all at hand if we wish. The future is open.

ACKNOWLEDGMENTS

I think, if you wrote a book, you fucked up.

—**Sam Bankman-Fried**

Many people made this book possible. But I am ultimately responsible for its contents. All errors, misrepresentations, and other inaccuracies are my own. And I know for a fact that at least a few of the following people disagree with at least some of the things I've written here.

This is all to say: None of this is their fault. Please don't be too hard on them.

* * *

Thanks first to my editor at Basic Books, T. J. Kelleher. He believed in this book when few others did, and steered me back on course when I nearly gave up hope myself. Thanks also to Lara Heimert, Liz Wetzel, Jen McArdle, Gillian Sutliff, Kristen Kim, Angie Messina, Alcimary Pena, Annie Chatham, Liz Dana, and the rest of the team at Basic and Hachette. And my agent, Peter Tallack at Curious Minds, has been unfailingly patient through the lengthy journey of this book to print, as has Louisa Pritchard, my foreign rights agent.

This book likely wouldn't exist at all if it weren't for the generosity of the Alfred P. Sloan Foundation. Thanks to Doron Weber for taking a gamble on this one, as well as Shriya Bhindwale and Ali Chunovic. Thanks also to the anonymous reviewers on my Sloan grant proposal, and the three manuscript reviewers for Sloan: Rediet Abebe, Meredith Broussard, and David Karpf.

That grant from Sloan allowed me to hire Caitlin Harrington, the most professional, thorough, and careful fact-checker I could have asked for. Only Caitlin could have corrected me about *Star Wars*. The Sloan grant also allowed me to hire the perfect researcher, John Wenz; I still don't understand how he knows the things he does. And after the incredible job Nick James did with the illustrations and photos for my last book, I knew I could count on him for this one.

The Simons Institute for the Theory of Computing at UC Berkeley provided me with an academic home and intellectual stimulation in the thick of my work on this book. Thanks to Kristin Kane and the rest of the team at Simons for giving me that opportunity, and to the participants in the 2023 Metacomplexity Program for being so welcoming and letting me play D&D and bar trivia with them.

At the end of my time working on this book, I was lucky enough to be a science journalism fellow at the Santa Fe Institute. Thanks to Katie Mast, Abha Eli, Carla Shedivy, David Feldman, David Krakauer, and the rest of the SFI team for an incredible five weeks. Thanks also to the rest of the participants in the 2024 SFI Complex Systems Summer School for a lovely and overwhelming month of science and art; thanks in particular to Anil Ananthaswamy for joining me on that wild adventure in the desert, and for his comments on part of this book.

In the very early days of my work on this book—really its prehistory—the UC Irvine Department of Logic and Philosophy of Science was kind enough to give me an academic affiliation. Thanks to Jeff Barrett for that, and for several helpful conversations and references early on.

More than anything, this book was made possible by conversations. Thanks to the people who took the time to talk with me on the record for this book—and to the people who talked with Caitlin during her fact-checking quest, who outnumbered my interviewees by at least a factor of two. In particular, thanks to Daniel Jordan, who has put up with many questions from me for many, many years; to David Thorstad, who was far more helpful and generous with his time than I could have reasonably hoped for; and to Adrian Daub, who gave me some excellent advice as I was first embarking on this project. Thanks also to Annalee Newitz, who walked all over the Mission with me while we commiserated about various distasteful subjects.

There are three people in particular who I can't imagine this book without: Adam Mann, Shannon Stirone, and John Wenz, the space humanists. When I look at this book, I see their influence everywhere. (Adam even added the second word to the title.) Our conversations built this thing, and their friendship and care made it possible for me to see it through to the end, despite the professional and personal tumult of the past few years. My dear friends Stefan Richter, Lisa Grossman, Nick James, and Leah Zani also helped keep my head above water in that rough surf. Megan Donahue kept me afloat too, and she helped me believe in my writing without the safety net of physics. (Megan also knows that the only true currency in this bankrupt world is what you share with someone else when you're uncool.) Finally, Megan and Adam took on the task of reading the entire manuscript before it went to print—I can't thank them enough for that.

There are many more people who deserve my thanks for their help with this book, including (but not limited to) Peter Aldhous, Christie Aschwanden, M. J. Bogatin, Brooke Borel, Marco Carmosino, Glen Chiacchieri, Katie Harmon Courage, Rachel Courtland, Ben Davis, Gaea Denker, Rose Eveleth, Adrienne Grant, Maureen Hanlon, Adam Hitchcock, Jon Jordan, Jenna Kanter, Kara Levy, Katie Mack, Maryn

McKenna, Andrew McNair, Randall Munroe, Will Rachelson, Ramin Skibba, Dana Smith, Christer Sturmark, Ariel Waldman, Zach Weinersmith, Marshall Weir, Alonzo Wickers, Tom Zeller, Ceila Zelman, and Carl Zimmer.

Thanks also to the Berkeley Public Library, the Oakland Public Library, and the University of California libraries. And perhaps most of all, thanks to the Mountain Lakes Public Library for starting me off right.

Finally, thanks to my parents and to the rest of my family. This one took a while, and I was pretty grumpy sometimes while I was working on it. Thanks for your patience, your understanding, and your love.

INTERVIEWS AND INTERVIEW REQUESTS

INTERVIEWS

David Chalmers, March 1, 2024, video call
Brad DeLong, December 21, 2023, Berkeley, CA
Jacob Foster, June 12, 2024, Santa Fe, NM
Timnit Gebru, August 23, 2023, Palo Alto, CA
David Gerard, May 27, 2023, London, UK
J. Storrs Hall, March 19, 2024, video call
Michael Hendricks, January 19, 2024, video call
Klaudia Jaźwińska, November 7, 2023, video call
Daniel Jordan, November 7 & 17, 2022, phone calls
Tsu-Jae King Liu, March 3, 2024, video call
Melanie Mitchell, August 30, 2023, video call
Toby Ord, July 10, 2023, video call
Rob Reich, February 12, 2024, video call
Benjamin Reinhardt, August 14, 2023, video call
Angela Saini, December 16, 2022, video call
Anders Sandberg, May 29, 2023, Oxford, UK
J. Fraser Stoddart, March 3, 2024, video call
Jaan Tallinn, June 13, 2023, video call
David Thorstad, April 26, 2023, video call; May 29, 2023, Oxford, UK
Émile Torres, July 18 & 26, 2023, video calls

Lucianne Walkowicz, February 27, 2020, Chicago, IL
Zach Weinersmith, August 24, 2023, video call
Şerife Wong, January 3, 2024, San Francisco, CA
Eliezer Yudkowsky, May 24, 2024, video call

INTERVIEW REQUESTS

Sam Altman (declined)
Marc Andreessen (declined)
Jeff Bezos (ignored)
Nick Bostrom (declined)
Eric Drexler (declined)
Ray Kurzweil (declined)
William MacAskill (canceled, ignored requests to reschedule)
Elon Musk (ignored)
Stuart Russell (declined)
Scott Siskind (declined)
Guillaume Verdon (ignored)

USAGE PERMISSIONS AND IMAGE CREDITS

“Torment Nexus” tweet © 2021 by Alex Blechman. Reprinted with the kind permission of Alex Blechman.

Excerpt from “Making Energy Too Cheap to Meter” © 2022 by Benjamin Reinhardt. Reprinted with the kind permission of Benjamin Reinhardt.

Figure 2.1 adapted from “File:PPTCountdowntoSingularityLog.jpg,” © Ray Kurzweil and Kurzweil Technologies, Inc. Licensed under a Creative Commons Attribution 1.0 Generic license (CC-BY-1.0). File has been modified from its original form. Original available on Wikimedia Commons: https://commons.wikimedia.org/wiki/File:PPTCountdowntoSingularityLog.jpg. License available at https://creativecommons.org/licenses/by/1.0/.

Figure 3.1 (a) image credit: NASA/JPL.

Figure 3.1 (b) image credit: NASA/JPL/Malin Space Science Systems.

Figure 5.1 (a) image credit: NASA/Rick Guidice.

Figure 5.1 (b) image credit: NASA.

NOTES

INTRODUCTION

1 Yudkowsky, interview with author.

2 Ibid.

3 Eliezer Yudkowsky, “Pausing AI Developments Isn’t Enough. We Need to Shut it All Down,” *Time*, March 29, 2023, https://time.com/6266923/ai-eliezer-yudkowsky-open-letter-not-enough/.

4 Sam Altman (@sama), Twitter (now X), February 3, 2023, https://twitter.com/sama/status/1621621725791404032.

5 Sam Altman, “Moore’s Law for Everything,” March 16, 2021, https://moores.samaltman.com/.

6 Cade Metz, “The ChatGPT King Isn’t Worried, but He Knows You Might Be,” *New York Times*, March 31, 2023, www.nytimes.com/2023/03/31/technology/sam-altman-open-ai-chatgpt.html.

7 Chamath Palihapitiya et al., “In Conversation with Sam Altman,” May 10, 2024, in *All-In*, podcast, YouTube, www.youtube.com/watch?v=nSM0xd8xHUM.

8 Elon Musk (@elonmusk), Twitter (now X), May 14, 2024, https://twitter.com/elonmusk/status/1790391774097088608.

9 Marc Andreessen, “The Techno-Optimist Manifesto,” Andreessen Horowitz, October 16, 2023, https://a16z.com/the-techno-optimist-manifesto/.

CHAPTER 1

1 Within limits. MacAskill, at least in his later writing, says that one shouldn’t take a job that is actively harmful, but he also claims that the harm inflicted by most lucrative jobs (e.g., a job in finance) is vastly outweighed by the good that could be done in the world if most of one’s salary from such a job were donated to worthy causes.

2 Adam Fisher, "Sam Bankman-Fried Has a Savior Complex—and Maybe You Should Too," Sequoia Capital, September 22, 2022, archived October 27, 2022, at the Wayback Machine, https://web.archive.org/web/20221027181005/https://www.sequoiacap.com/article/sam-bankman-fried-spotlight/; Nicholas Kulish, "How a Scottish Moral Philosopher Got Elon Musk's Number," *New York Times*, October 8, 2022, www.nytimes.com/2022/10/08/business/effective-altruism-elon-musk.html; Zeke Faux, *Number Go Up* (New York: Currency Books, 2023), 83; Gideon Lewis-Kraus, "The Reluctant Prophet of Effective Altruism," *New Yorker*, August 8, 2022, www.newyorker.com/magazine/2022/08/15/the-reluctant-prophet-of-effective-altruism.

3 William MacAskill (@willmacaskill), Twitter, November 11, 2022, https://twitter.com/willmacaskill/status/1591218022362284034.

4 Peter Singer, "Famine, Affluence, and Morality," *Philosophy & Public Affairs* 1, no. 3 (1972): 229–243, www.jstor.org/stable/2265052; Peter Singer, *The Life You Can Save* (New York: Random House, 2009), 18.

5 Peter Singer, "The Drowning Child and the Expanding Circle," *New Internationalist*, April 5, 1997, https://newint.org/features/1997/04/05/peter-singer-drowning-child-new-internationalist.

6 Robert Wiblin, "Most People Report Believing It's Incredibly Cheap to Save Lives in the Developing World," 80,000 Hours, May 9, 2017, https://80000hours.org/2017/05/most-people-report-believing-its-incredibly-cheap-to-save-lives-in-the-developing-world/; GiveWell, "How We Produce Impact Estimates," last modified February 2024, www.givewell.org/impact-estimates#Impact_metrics_for_grants_to_GiveWells_top_charities.

7 Thomas Hurka, "Moral Demands and Permissions/Prerogatives," *Stanford Encyclopedia of Philosophy* (Fall 2024), https://plato.stanford.edu/entries/moral-demands-permissions/; Travis Timmerman, "Sometimes There Is Nothing Wrong with Letting a Child Drown," *Analysis* 75, no. 2 (2015): 204–212, www.jstor.org/stable/24671242.

8 Lewis-Kraus, "Reluctant Prophet."

9 Rob Mather, "Against Malaria Foundation: What We Do, How We Do It, and the Challenges," Effective Altruism, accessed June 12, 2024, www.effectivealtruism.org/articles/ea-global-2018-amf-rob-mather.

10 Lewis-Kraus, "Reluctant Prophet."

11 "About Us: What Do We Do, and How Can We Help?," 80,000 Hours, accessed June 12, 2024, https://80000hours.org/about/.

12 William MacAskill, "About William MacAskill," accessed June 12, 2024, www.williammacaskill.com/press.

13 Naina Bajekal, "Want to Do More Good? This Movement Might Have the Answer," *Time*, August 10, 2022, https://time.com/6204627/effective-altruism-longtermism-william-macaskill-interview/.

14 "History," Centre for Effective Altruism, accessed June 12, 2024, www.centreforeffectivealtruism.org/history; "Oxford-Based Charity Receives More Than $2.5 Billion in Pledges from 'Community of Effective Givers,'" news release, University of Oxford,

March 1, 2022, www.ox.ac.uk/news/2022-03-01-oxford-based-charity-receives-more-25-billion-pledges-community-effective-givers.

15 William MacAskill, *What We Owe the Future* (New York: Basic Books, 2022), 9.

16 Ibid., 5.

17 Ibid., 10.

18 Ibid., 13.

19 Ibid., 5, 27–28.

20 This calculation assumes that the mean lifespan over that time is one hundred years and actually uses a population figure of ten billion, just to make the math easier. Lowering the mean lifespan to a more realistic figure of seventy years and lowering the population figure to eight billion people—i.e., making things more like the actual world we live in today—would actually result in a somewhat higher estimate for the total number of future humans, but the order of magnitude remains the same.

21 There are two trillion galaxies in the observable universe: NASA Hubble Mission Team, "Hubble Reveals Observable Universe Contains 10 Times More Galaxies Than Previously Thought," NASA, October 13, 2016, www.nasa.gov/feature/goddard/2016/hubble-reveals-observable-universe-contains-10-times-more-galaxies-than-previously-thought. But most of them lie outside the Hubble volume, the maximum distance we can travel to: Pim van Oirschot, Juliana Kwan, and Geraint F. Lewis, "Through the Looking Glass: Why the 'Cosmic Horizon' Is Not a Horizon," *Monthly Notices of the Royal Astronomical Society* 404, no. 4 (June 2010): 1633–1638, https://doi.org/10.1111/j.1365-2966.2010.16398.x. Given current estimates of the Hubble parameter and the size of the observable universe, this means that approximately 3 percent of the observable universe is currently accessible. So that leaves us with between 10^{10} and 10^{11} accessible galaxies; we'll use the lower number as a conservative estimate. Most stars are M dwarfs, and most of them will die (i.e., stop fusing nuclei in their cores) in about 10^{13} years: Fred C. Adams and Gregory Laughlin, "A Dying Universe: The Long-Term Fate and Evolution of Astrophysical Objects," *Reviews of Modern Physics* 69, no. 337 (April 1, 1997), https://doi.org/10.1103/RevModPhys.69.337. Combining these figures with the estimate of 10^{26} people per billion years per galaxy developed in the main text, we get an answer of 10^{40} future people in this scenario.

22 MacAskill, *What We Owe*, 189. (I had hoped that this was an intentional *Star Trek* reference, but alas—MacAskill confirmed it's not.)

23 MacAskill, *What We Owe*, 253.

24 Hilary Greaves and William MacAskill, "The Case for Strong Longtermism" (GPI Working Paper No. 5-2021, Global Priorities Institute, June 2021), https://globalprioritiesinstitute.org/wp-content/uploads/The-Case-for-Strong-Longtermism-GPI-Working-Paper-June-2021-2-2.pdf.

25 Lewis-Kraus, "Reluctant Prophet."

26 Ezra Klein, "Three Sentences That Could Change the World—and Your Life," August 9, 2022, in *The Ezra Klein Show*, podcast, transcript, *New York Times*, www.nytimes.com/2022/08/09/opinion/ezra-klein-podcast-will-macaskill.html.

27 Bajekal, "Want to Do More Good?"

28 "Hardcover Nonfiction Best Sellers," *New York Times*, September 18, 2022, www.nytimes.com/books/best-sellers/2022/09/18/hardcover-nonfiction/.

29 Blurb appears on the Amazon page for *What We Owe the Future*: "What We Owe the Future," Amazon, accessed March 24, 2023, www.amazon.com/What-Owe-Future-William-MacAskill/dp/1541618629.

30 Mixed and negative reviews: Regina Rini, "An Effective Altruist?," *Times Literary Supplement*, September 9, 2022, www.the-tls.co.uk/articles/what-we-owe-the-future-william-macaskill-book-review-regina-rini/; Émile P. Torres, "Understanding 'Longtermism': Why This Suddenly Influential Philosophy Is So Toxic," *Salon*, August 20, 2022, www.salon.com/2022/08/20/understanding-longtermism-why-this-suddenly-influential-philosophy-is-so/; Barton Swaim, "'What We Owe the Future' Review: A Technocrat's Tomorrow," *Wall Street Journal*, August 26, 2022, www.wsj.com/articles/what-we-owe-the-future-review-a-technocrats-tomorrow-11661544593; Kieran Setiya, "The New Moral Mathematics," *Boston Review*, August 15, 2022, www.bostonreview.net/articles/the-new-moral-mathematics/. See also Leif Wenar, "The Deaths of Effective Altruism," *Wired*, March 27, 2024, https://www.wired.com/story/deaths-of-effective-altruism/.

31 Ariana Eunjung Cha, "Cari Tuna and Dustin Moskovitz: Young Silicon Valley Billionaires Pioneer New Approach to Philanthropy," *Washington Post*, December 26, 2014, www.washingtonpost.com/business/billionaire-couple-give-plenty-to-charity-but-they-do-quite-a-bit-of-homework/2014/12/26/19fae34c-86d6-11e4-b9b7-b8632ae73d25_story.html.

32 "About Us," Open Philanthropy, accessed March 24, 2023, www.openphilanthropy.org/about-us/. Donation data pulled from the full database of Open Philanthropy grants, available as a CSV here: www.openphilanthropy.org/wp-admin/admin-ajax.php?action=generate_grants&nonce=79a8b1a618 (accessed August 30, 2024). List of Effective Ventures constituent organizations found on their homepage, EV.org, accessed August 30, 2024.

33 Gian Volpicelli, "Stop the Killer Robots! Musk-Backed Lobbyists Fight to Save Europe from Bad AI," *Politico*, November 24, 2022, www.politico.eu/article/meet-the-musk-backed-ngos-trying-to-save-europe-from-bad-ai/; "Future of Life Institute," Effective Altruism Forum, accessed August 30, 2022, https://forum.effectivealtruism.org/topics/future-of-life-institute; "Our Mission," Future of Life Institute, accessed August 30, 2024, https://futureoflife.org/our-mission/.

34 Future of Life Institute, accessed June 12, 2024, https://futureoflife.org/; Max Tegmark, "Elon Musk Donates $10M to Keep AI Beneficial," Future of Life Institute, January 15, 2015, archived November 22, 2019, at the Wayback Machine, https://web.archive.org/web/20191122175152/https://futureoflife.org/2015/10/12/elon-musk-donates-10m-to-keep-ai-beneficial/; "Future of Life Institute," EU Transparency Register, archived May 26, 2023, at the Wayback Machine, https://web.archive.org/web/20230526011529/https:/ec.europa.eu/transparencyregister/public/consultation/displaylobbyist.do?id=787064543128-10; Future of Life Institute, private communication.

35 Brendan Bordelon, "The Little-Known AI Group That Got $660 Million," *Politico*, March 26, 2024, www.politico.com/news/2024/03/25/a-665m-crypto-war-chest-roils-ai-safety-fight-00148621; Future of Life Institute, private communication. See also Future of Life Institute's 2021 Form 990, available on ProPublica's Nonprofit Explorer: https://projects.propublica.org/nonprofits/organizations/471052538/202400669349300240/full.

36 MacAskill points out that the futures he's talking about wouldn't have truly unlimited growth, but more on that later.

37 Christianna Reedy, "Kurzweil Claims That the Singularity Will Happen by 2045," *Futurism*, October 16, 2017, https://futurism.com/kurzweil-claims-that-the-singularity-will-happen-by-2045.

38 David Kushner, "When Humans and Machines Merge," *Rolling Stone*, February 19, 2009, www.rollingstone.com/culture/culture-features/ray-kurzweil-ai-david-kushner-1234779424/.

39 Mathias Döpfner, "Jeff Bezos Interview with Axel Springer CEO on Amazon, Blue Origin, Family," *Business Insider*, April 28, 2018, www.businessinsider.com/jeff-bezos-interview-axel-springer-ceo-amazon-trump-blue-origin-family-regulation-washington-post-2018-4.

40 Dave Mosher, "Jeff Bezos Just Gave a Private Talk in New York. From Utopian Space Colonies to Dissing Elon Musk's Martian Dream, Here Are the Most Notable Things He Said," *Business Insider*, February 23, 2019, www.businessinsider.com/jeff-bezos-blue-origin-wings-club-presentation-transcript-2019-2.

41 Elon Musk (@elonmusk), Twitter (now X), June 24, 2018, https://twitter.com/elonmusk/status/1011083630301536256; see also, Nick Lucchesi, "Elon Musk Calls on the Public to 'Preserve Human Consciousness' with Starship," *Inverse*, September 28, 2019, www.inverse.com/article/59676-spacex-starship-presentation.

42 Richard Fuisz et al., "New World: Interstellar," February 22, 2019, in *Anatomy of Next*, podcast, Founders Fund, foundersfund.com/2019/02/anatomy-next-season-2-episode-11/; "Michael Solana," Founders Fund, accessed August 30, 2024, foundersfund.com/team/michael-solana/.

43 Thomas W. Murphy, 2022, "Limits to Economic Growth," *Nature Physics* 18, no. 8: 844–847, https://doi.org/10.1038/s41567-022-01652-6; Thomas Murphy, private communication.

44 This is all ignoring the fact that at light speed we can't even reach all the stars in the Milky Way in that amount of time, and can never reach most of the stars in the observable universe at all. So this analysis is actually far too generous—realistically, growth must end far earlier than that.

45 Tom Murphy, "Galactic-Scale Energy," *Do the Math* (blog), July 12, 2011, https://dothemath.ucsd.edu/2011/07/galactic-scale-energy/; Tom Murphy, "Can Economic Growth Last?," *Do the Math* (blog), July 14, 2011, https://dothemath.ucsd.edu/2011/07/can-economic-growth-last/. The figure for the observable universe assumes there are two trillion galaxies visible, each containing roughly as many stars as the Milky Way.

This is a deliberately large estimate, and most of those galaxies are inaccessible anyhow, so once again, this is a generous figure for the ultimate limit.

46 MacAskill, *What We Owe*, 145.

47 Ibid., 144, 162.

48 Peter Singer, "The Hinge of History," *Project Syndicate*, October 8, 2021, www.project-syndicate.org/commentary/ethical-implications-of-focusing-on-extinction-risk-by-peter-singer-2021-10.

49 William MacAskill (@willmacaskill), Twitter (now X), November 11, 2022, https://twitter.com/willmacaskill/status/1591218030364995585. Note that in the next tweets in the thread, MacAskill cites himself and two other prominent longtermists (Toby Ord and Holden Karnofsky) saying things to this effect.

50 Elon Musk (@elonmusk), Twitter (now X), August 1, 2022, https://twitter.com/elonmusk/status/1554335028313718784.

51 Marisa Taylor, "At SpaceX, Worker Injuries Soar in Elon Musk's Rush to Mars," Reuters, November 10, 2023, www.reuters.com/investigates/special-report/spacex-musk-safety/.

52 Leonard David, "Jeff Bezos' Vision: 'A Trillion Humans in the Solar System,'" Space.com, July 21, 2017, www.space.com/37572-jeff-bezos-trillion-people-solar-system.html.

53 Eliezer Yudkowsky (@ESYudkowsky), Twitter (now X), December 31, 2018, https://twitter.com/ESYudkowsky/status/1079644135907151872.

54 Dylan Matthews, "You Have 80,000 Hours in Your Career. Here's How to Do the Most Good with Them," *Vox*, August 3, 2015, www.vox.com/2015/7/29/9067641/william-macaskill-effective-altruism.

55 William MacAskill, "Replaceability, Career Choice, and Making a Difference," *Ethical Theory and Moral Practice* 17 (2014): 269–283, https://doi.org/10.1007/s10677-013-9433-4.

56 William MacAskill, personal communication.

57 "Sam Bankman-Fried," 80,000 Hours, archived June 13, 2021, at the Wayback Machine, https://web.archive.org/web/20210613111013/https://80000hours.org/stories/sam-bankman-fried/.

58 Fisher, "Sam Bankman-Fried Has a Savior Complex."

59 Reed Albergotti and Liz Hoffman, "Charity-Linked Money Launched Sam Bankman-Fried's Empire," *Semafor*, December 8, 2022, www.semafor.com/article/12/07/2022/charity-money-launched-sam-bankman-frieds-empire.

60 White House, "FACT SHEET: Climate and Energy Implications of Crypto-Assets in the United States," news release, September 8, 2022, www.whitehouse.gov/ostp/news-updates/2022/09/08/fact-sheet-climate-and-energy-implications-of-crypto-assets-in-the-united-states/; Hannah Ritchie, Pablo Rosado, and Max Roser, "CO2 and Greenhouse Gas Emissions," database, Our World in Data, accessed March 27, 2023, https://ourworldindata.org/co2-and-greenhouse-gas-emissions.

61 "Crypto Firm FTX Trading's Valuation Rises to $18 bln After $900 mln Investment," Reuters, July 20, 2021, www.reuters.com/technology/crypto-firm-ftx-trading-raises-900-mln-18-bln-valuation-2021-07-20/.

62 Bill Chappell, David Gura, and Lisa Lambert, "Bankman-Fried Is Arrested as Feds Charge Massive Fraud at FTX Crypto Exchange," NPR, December 13, 2022, www.npr.org/2022/12/12/1142361088/bankman-fried-ceo-ftx-crypto-exchange-arrested-bahamas-charges-sdny; David Gura, "Sam Bankman-Fried Is Found Guilty of All Charges in FTX's Spectacular Collapse," NPR, November 2, 2023, www.npr.org/2023/11/02/1210100678/sam-bankman-fried-trial-verdict-ftx-crypto; Rafael Nam, "Sam Bankman-Fried Sentenced to 25 Years in Prison for His FTX Crimes," NPR, March 28, 2024, www.npr.org/2024/03/28/1241210300/sam-bankman-fried-ftx-sentencing-crimes-crypto-mogul-greed.

63 Zachary Robinson, "EV Updates: FTX Settlement and the Future of EV," Effective Altruism Forum, December 13, 2023, https://forum.effectivealtruism.org/posts/HjsfHwqasyQMWRzZN/ev-updates-ftx-settlement-and-the-future-of-ev.

64 "Who We Are," Future Fund, archived November 9, 2022, at the Wayback Machine, https://web.archive.org/web/20221109230958/https://ftxfuturefund.org/about/; "Our Grants and Investments," Future Fund, archived November 9, 2022, at the Wayback Machine, https://web.archive.org/web/20221109231205/https://ftxfuturefund.org/our-grants/.

65 "Support Future Perfect," *Vox*, August 18, 2022, www.vox.com/2020/1/7/21020439/support-future-perfect; Katie Clinebell, "By Returning $10M, Semafor Becomes the Latest Media Outlet Distancing Itself from SBF," *Investopedia*, January 18, 2023, www.investopedia.com/media-outlets-are-returning-sam-bankman-fried-s-funds-7096408; "ProPublica Returns Grant Funded by Bankman-Fried Family," ProPublica, February 28, 2022, updated December 20, 2022, www.propublica.org/atpropublica/bankman-fried-family-donates-5-million-to-propublica.

66 Dylan Matthews, "How Effective Altruism Let Sam Bankman-Fried Happen," *Vox*, December 12, 2022, www.vox.com/future-perfect/23500014/effective-altruism-sam-bankman-fried-ftx-crypto.

67 Jennifer Szalai, "How Sam Bankman-Fried Put Effective Altruism on the Defensive," *New York Times*, December 9, 2022, www.nytimes.com/2022/12/09/books/review/effective-altruism-sam-bankman-fried-crypto.html; Annie Lowrey, "Effective Altruism Committed the Sin It Was Supposed to Correct," *The Atlantic*, November 17, 2022, www.theatlantic.com/ideas/archive/2022/11/cryptocurrency-effective-altruism-ftx-sam-bankman-fried/672149/; Gideon Lewis-Kraus, "Sam Bankman-Fried, Effective Altruism, and the Question of Complicity," *New Yorker*, December 1, 2022, www.newyorker.com/news/annals-of-inquiry/sam-bankman-fried-effective-altruism-and-the-question-of-complicity; Nitasha Tiku, "The Do-Gooder Movement That Shielded Sam Bankman-Fried from Scrutiny," *Washington Post*, November 17, 2022, www.washingtonpost.com/technology/2022/11/17/effective-altruism-sam-bankman-fried-ftx-crypto/; Eric Levitz, "Is Effective Altruism to Blame for Sam Bankman-Fried?," *New York*, November 16, 2022, https://nymag.com/intelligencer/2022/11/effective-altruism-sam-bankman-fried-sbf-ftx-crypto.html; Zeeshan Aleem, "How Sam Bankman-Fried's

Fall Exposes the Perils of Effective Altruism," MSNBC, December 3, 2022, www.msnbc.com/opinion/msnbc-opinion/ftx-sbf-effective-altruism-bankman-fried-rcna59172.

68 David Yaffe-Bellany, Lora Kelley, and Cade Metz, "She Was a Little-Known Crypto Trader. Then FTX Collapsed," *New York Times*, November 23, 2022, www.nytimes.com/2022/11/23/business/ftx-caroline-ellison-sbf.html.

69 Tarpley Hitt, "Are These Caroline Ellison's Tumblrs?," *Gawker*, November 14, 2022, archived March 28, 2023, at the Wayback Machine, https://web.archive.org/web/20230328082349/gawker.com/money/are-these-caroline-ellisons-tumblrs.

70 "worldoptimization," archived November 11, 2022, accessed March 28, 2023, https://caroline.milkyeggs.com/worldoptimization. Formatting in the original.

71 *Harry Potter and the Methods of Rationality*, by one Eliezer Yudkowsky. Less Wrong [Eliezer Yudkowsky], *Harry Potter and the Methods of Rationality*, serialized novel, FanFiction.net, February 28, 2010, www.fanfiction.net/s/5782108/1/Harry_Potter_and_the_Methods_of_Rationality.

72 Kelsey Piper, "Sam Bankman-Fried Tries to Explain Himself," *Vox*, November 16, 2022, www.vox.com/future-perfect/23462333/sam-bankman-fried-ftx-cryptocurrency-effective-altruism-crypto-bahamas-philanthropy.

73 "Transcript of Sam Bankman-Fried's Interview at the DealBook Summit," *New York Times*, December 1, 2022, www.nytimes.com/2022/12/01/business/dealbook/sam-bankman-fried-dealbook-interview-transcript.html.

74 Lucianne Walkowicz, interview with the author.

CHAPTER 2

1 Kushner, "When Humans and Machines Merge."

2 "Futurist Ray Kurzweil Says He Can Bring His Dead Father Back to Life Through a Computer Avatar," ABC News, August 9, 2011, https://abcnews.go.com/Technology/futurist-ray-kurzweil-bring-dead-father-back-life/story?id=14267712.

3 Peter Kirn, "That Time in 1965 When a Teen Ray Kurzweil Made a Computer Compose Music and Met LBJ," *CDM*, March 3, 2020, https://cdm.link/2020/03/ray-kurzweil-ai-music-1965/; "Raymond Kurzweil," Lemelson-MIT, accessed August 31, 2024, https://lemelson.mit.edu/resources/raymond-kurzweil; interview with Ray Kurzweil conducted by Dag Spicer on behalf of the Computer History Museum, July 13, 2009, Mountain View, CA, CHM Reference number X5461.2010, https://archive.computerhistory.org/resources/access/text/2013/05/102702108-05-01-acc.pdf.

4 "Fredric Kurzweil, Music Educator, 57," *New York Times*, August 13, 1970, https://timesmachine.nytimes.com/timesmachine/1970/08/13/90616871.html.

5 Kushner, "When Humans and Machines Merge."

6 Ray Kurzweil, *The Age of Intelligent Machines* (Cambridge, MA: MIT Press, 1990), 133.

7 AP, "Gates: Get Ready for Chip Implants," CNN, July 5, 2005, archived July 8, 2005, at the Wayback Machine, https://web.archive.org/web/20050708012222/http://edition.cnn.com/2005/TECH/07/04/gates.implants.ap/.

8 Ray Kurzweil, *How My Predictions Are Faring*, October 2010, www.thekurzweillibrary.com/images/How-My-Predictions-Are-Faring.pdf. For some independent estimates, see Alex Knapp, "Ray Kurzweil's Predictions for 2009 Were Mostly Inaccurate," *Forbes*, June 2, 2013, www.forbes.com/sites/alexknapp/2012/03/20/ray-kurzweils-predictions-for-2009-were-mostly-inaccurate/; Daniel Lyons, "I, Robot," *Newsweek*, May 16, 2009, archived April 13, 2010, at the Wayback Machine, https://web.archive.org/web/20100413081045/http://www.newsweek.com/id/197812; Dan Luu, "Futurist Prediction Methods and Accuracy," September 2022, https://danluu.com/futurist-predictions/.

9 This specific quote comes from Kurzweil's 2024 South by Southwest interview: Ray Kurzweil, "Featured Session: The Singularity Is Nearer" (South by Southwest, Austin, TX, March 10, 2024), https://schedule.sxsw.com/2024/events/PP1143806. He's also made the same or substantially similar claims since at least 2002, when he made a bet with Mitch Kapor: Mitchell Kapor and Raymond Kurzweil, "By 2029 No Computer—or 'Machine Intelligence'—Will Have Passed the Turing Test," Long Bets, accessed June 13, 2024, https://longbets.org/1/. He also said the same thing in 2005: Ray Kurzweil, *The Singularity Is Near: When Humans Transcend Biology* (New York: Viking, 2005). And in 2017, in this interview: Christianna Reedy, "Kurzweil Claims That the Singularity Will Happen by 2045," *Futurism*, October 16, 2017, https://futurism.com/kurzweil-claims-that-the-singularity-will-happen-by-2045.

10 Specific quote is from Reedy, "Kurzweil Claims." But he makes the same claim on the timing of the Singularity in Kurzweil, *Singularity Is Near*, and in Kurzweil, "Featured Session."

11 Kurzweil, *Singularity Is Near*, 12.

12 Kushner, "When Humans and Machines Merge."

13 Reedy, "Kurzweil Claims."

14 Kurzweil, *Singularity Is Near*, 9.

15 Kushner, "When Humans and Machines Merge."

16 Kurzweil, "Featured Session."

17 Kushner, "When Humans and Machines Merge."

18 Gordon E. Moore, "Cramming More Components onto Integrated Circuits," *Electronics* 38, no. 8 (April 19, 1965), www.computerhistory.org/collections/catalog/102770822; Rachel Courtland, "Gordon Moore: The Man Whose Name Means Progress," *IEEE Spectrum*, March 30, 2015, https://spectrum.ieee.org/gordon-moore-the-man-whose-name-means-progress; Michael Kanellos, "Moore's Law to Roll on for Another Decade," CNET News, February 10, 2003, archived July 19, 2011, at the Wayback Machine, https://web.archive.org/web/20110719150309/http://news.cnet.com/2100-1001-984051.html.

19 Kurzweil, *Singularity Is Near*, 40.

20 Ibid., 12.

21 Ibid., 47.

22 Ibid., 18–19.

23 Ibid., 47.

24 Ibid., 260.

25 Ibid., 145.

26 Ibid., 28.

27 Ibid., 20, 24.

28 Ibid., 375.

29 I. J. Good, ed., *The Scientist Speculates: An Anthology of Partly-Baked Ideas* (New York: Basic Books, 1962), 1. Emphasis on "pbi" taken from the identical quote on the back dust jacket.

30 Ibid., on the back dust-jacket flap under "About the Editor."

31 One of the essays in the volume, by the noted physicist Eugene Wigner, was the first published article to discuss the quantum thought experiment that became known as "Wigner's friend."

32 Hugh Sebag-Montefiore, *Enigma: The Battle for the Code* (Hoboken, NJ: John Wiley & Sons, 2000), 192.

33 I. J. Good, "Review: *Calculating Instruments and Machines* by D. R. Hartree," *Journal of the Royal Statistical Society* 114, no. 1 (1951): 106–107, www.jstor.org/stable/2980914.

34 Good, *Scientist Speculates*, 194–195.

35 I. J. Good, "Speculations Concerning the First Ultraintelligent Machine," in *Advances in Computers*, vol. 6, eds. Franz L. Alt and Morris Rubinoff (New York: Academic Press, 1965) (emphasis his).

36 S. Ulam, "John von Neumann, 1903–1957," *Bulletin of the American Mathematical Society* 64, no. 3, part 2 (May 1958), www.ams.org/journals/bull/1958-64-03/S0002-9904-1958-10189-5/S0002-9904-1958-10189-5.pdf.

37 He had first written about the idea of the Singularity in a short opinion piece in *Omni* magazine three years earlier: Vernor Vinge, "First Word," *Omni*, January 1983.

38 Vernor Vinge, *Marooned in Realtime* (New York: Baen Books, 1986), 313.

39 Vernor Vinge, "The Coming Technological Singularity: How to Survive in the Post-Human Era" (paper presented at Vision 21: Interdisciplinary Science and Engineering in the Era of Cyberspace, NASA Lewis Research Center, Cleveland, OH, March 30–31, 1993), https://edoras.sdsu.edu/~vinge/misc/singularity.html.

40 James Barrat, *Our Final Invention: Artificial Intelligence and the End of the Human Era* (New York: Thomas Dunne, 2013), 121.

41 Ed Regis, "Meet the Extropians," *Wired*, October 1, 1994, www.wired.com/1994/10/extropians/.

42 Nick Bostrom, "A History of Transhumanist Thought," *Journal of Evolution and Technology* 14, no. 1 (April 2005), https://jetpress.org/volume14/bostrom.pdf; *Extropy: Vaccine for Future Shock*, no. 1 (Fall 1988), 2, https://github.com/Extropians/Extropy/blob/master/Extropy-01.pdf. See also, H+Pedia, s.v. "Extropy Magazines," last modified November 30, 2023, 01:06, https://hpluspedia.org/wiki/Extropy_Magazines.

43 *Extropy* no. 1, 3.

44 Bostrom, "History of Transhumanist Thought."

45 Ibid.

46 Parties: Regis, "Meet the Extropians"; Jim McClellan, "The Tomorrow People," *UK Observer*, March 26, 1995, https://mason.gmu.edu/~rhanson/press/UKObserver-3-26-95.htm; Jon Evans, "Extropia's Children, Chapter 1," *Gradient Ascendant* (blog), October 17, 2022, https://aiascendant.substack.com/p/extropias-children-chapter-1-the-wunderkind. Conferences and MIT student org: David Thorstad, "Belonging (Part 1: That Bostrom Email)," *Reflective Altruism* (blog), January 12, 2023, https://reflectivealtruism.com/2023/01/12/off-series-that-bostrom-email/.

47 Ed Regis, *Great Mambo Chicken and the Transhuman Condition* (New York: Basic Books, 1990); and Regis, "Meet the Extropians," for example.

48 Evans, "Extropia's Children"; Meghan O'Gieblyn, "God in the Machine: My Strange Journey into Transhumanism," *The Guardian*, April 18, 2017, www.theguardian.com/technology/2017/apr/18/god-in-the-machine-my-strange-journey-into-transhumanism; Ben Goertzel, "The Extropian Creed," September 2000, accessed June 13, 2024, www.goertzel.org/benzine/extropians.htm.

49 Kurzweil, *Singularity Is Near*, 369.

50 Vinge, "Coming Technological Singularity" (ellipsis in original).

51 Ken MacLeod, introduction to *The Atrocity Archives*, by Charles Stross (New York: Ace Books, 2009), xix.

52 Jo Walton, *What Makes This Book So Great: Re-reading the Classics of Science Fiction and Fantasy* (New York: Tor Books, 2014), 26 (emphasis in original).

53 Joshua Thomas Raulerson, "Singularities: Technoculture, Transhumanism, and Science Fiction in the 21st Century" (doctoral diss., University of Iowa, Spring 2010), https://doi.org/10.17077/etd.w52umc47. See also Brooks Landon, "That Light at the End of the Tunnel: The Plurality of Singularity," *Science Fiction Studies* 29, no. 1 (March 2012), www.jstor.org/stable/10.5621/sciefictstud.39.1.0002.

54 Good certainly knew this. In his 1965 paper, the quote above continues with the following: "It is curious that this point is made so seldom outside of science fiction. It is sometimes worthwhile to take science fiction seriously." Good, "Speculations Concerning."

55 Isaac Asimov, *I, Robot* (New York: Bantam Doubleday, 1950), 246.

56 John W. Campbell Jr., "The Last Evolution," *Amazing Stories*, August 1932, www.gutenberg.org/files/27462/27462-h/27462-h.htm.

57 Asimov, *I, Robot*, copyright page. The dedication to Campbell is at the top of the page.

58 Asimov, *I, Robot*, 271 (emphasis his).

59 Kurzweil, *Singularity Is Near*, 14.

60 NASA, "NASA Ames Becomes Home to Newly Launched Singularity University," news release, February 3, 2009, www.nasa.gov/news-release/nasa-ames-becomes-home-to-newly-launched-singularity-university/; Nokia Research Center, "Nokia Supports Singularity University as Fifth Corporate Founder," news release, archived January 27, 2013, at the Wayback Machine, https://web.archive.org/web/20130127224419/http://research.nokia.com/news/11357.

61 David J. Hill, "Exclusive Interview: Ray Kurzweil Discusses His First Two Months at Google," *Singularity Hub*, March 19, 2013, https://singularityhub.com/2013/03/19/exclusive-interview-ray-kurzweil-discusses-his-first-two-months-at-google/.

62 Jaron Lanier, "The First Church of Robotics," *New York Times*, August 9, 2010, www.nytimes.com/2010/08/09/opinion/09lanier.html.

63 "The Singularity Is Nearer: When We Merge with AI," Amazon, accessed June 13, 2024, www.amazon.com/Singularity-Nearer-Ray-Kurzweil-ebook/dp/B08Y6FYJVY.

64 Dylan Matthews, "This Oxford Professor Thinks Artificial Intelligence Will Destroy Us All," *Vox*, August 19, 2014, www.vox.com/2014/8/19/6031367/oxford-nick-bostrom-artificial-intelligence-superintelligence.

65 David J. Chalmers, "The Singularity: A Philosophical Analysis," *Journal of Consciousness Studies* 17 (2010): 3.

66 Chalmers, interview with author.

67 Chuck Klosterman, *But What If We're Wrong?* (New York: Penguin, 2016), 156.

68 Erin Blakemore, "Think Customer Service Is Bad Now? Read This 4,000-Year-Old Complaint Letter," *National Geographic*, May 31, 2024, www.nationalgeographic.com/history/article/ea-nasir-copper-merchant-ur; "Heofonum || dado" (@Heofonum), Twitter (now X), August 30, 2021, https://twitter.com/Heofonum/status/1432446141329772556.

69 Liu, interview with author.

70 Tsu-Jae King Liu, personal communication, March 4, 2024 (emphasis mine).

71 Ibid.

72 Manek Dubash, "Moore's Law Is Dead, Says Gordon Moore," *Computer World*, April 13, 2010, archived June 13, 2020, at the Wayback Machine, https://web.archive.org/web/20200613232824/https://www.computerworld.com/article/3554889/moore-s-law-is-dead-says-gordon-moore.html.

73 Liu interview.

74 Nicholas Bloom et al., "Are Ideas Getting Harder to Find?," *American Economic Review* 110, no. 4 (2020): 1104–1044, https://doi.org/10.1257/aer.20180338 (emphasis theirs). Found via David Thorstad, "Against the Singularity Hypothesis" (GPI working paper No. 19-2022, Global Priorities Institute, 2022), https://globalprioritiesinstitute

.org/against-the-singularity-hypothesis-david-thorstad/, and subsequent interviews with Thorstad.

75 Kurzweil, *Singularity Is Near*, 8. It's worth noting that here and in several other places in the book, Kurzweil talks about exponential curves as though they have an inflection point—he mentions the "knee of the curve" on p. 10, for example. They don't. Zooming in on an exponential curve shows that it looks the same at all points; the growth rate is constant. That's why exponential curves are straight lines on the logarithmic plots Kurzweil is so fond of. This is a puzzling mistake from Kurzweil, and it's hard to imagine he doesn't know it's a mistake.

76 There is one possible exception here, which is the exponential expansion of a universe dominated by dark energy, as ours appears to be. (There is also the related cosmological phenomenon of "eternal inflation.") But even the exponential expansion of our universe may end someday with vacuum decay. Regardless of that, the exponential expansion of the universe is actually bad news for Kurzweil's program. For more on this, see Chapter 4.

77 Heinz von Foerster, Patricia M. Mora, and Lawrence W. Amiot, "Doomsday: Friday, 13 November, A.D. 2026," *Science* 132 (1960): 1291–1295, https://doi.org/10.1126/science.132.3436.1291.

78 Kurzweil, *Singularity Is Near*, 44.

79 Ibid., 66–67.

80 Ibid., 42–43.

81 Bloom et al., "Are Ideas Getting Harder."

82 David Rotman, "We're Not Prepared for the End of Moore's Law," *MIT Technology Review*, February 24, 2020, www.technologyreview.com/2020/02/24/905789/were-not-prepared-for-the-end-of-moores-law/.

83 Thorstad, interview with author, April 26, 2023.

84 Neil C. Thompson, Shuning Ge, and Gabriel F. Manso, "The Importance of (Exponentially More) Computing Power," arXiv:2206.14007 (2022), https://doi.org/10.48550/arXiv.2206.14007, via Thorstad, "Against the Singularity Hypothesis."

85 Cosma Shalizi, "g, a Statistical Myth," *Three-Toed Sloth* (blog), October 18, 2007, http://bactra.org/weblog/523.html.

86 Following Chalmers in his paper on the Singularity, we could instead talk about an ability explosion, in which the ability of a computer to perform a well-defined task (one with a clear metric) increases exponentially. That ability would probably have to be designing a computer with even greater ability (or something closely correlated with that), otherwise it's not clear why there would be an explosion in the first place. Yet that ability seems like a rather narrow foundation to serve as the basis for a singularity, as it's more than slightly circular. While, as Chalmers says, computer-designing ability is probably associated with other abilities of interest, like mathematical ability, it's not at all clear that an exponential increase in the ability of a computer to, say, do mathematics would lead to a fundamental change in the technological progress of the world. And this all assumes that such an exponential increase would happen. Given the other

possible barriers in the way (the end of Moore's law, the nonlinear relationship between intelligence and computing power, the possible unsuitability of computers as machines to simulate the human brain, and so on), the case for a mere ability explosion in a few narrow domains still looks rather weak.

87 Kurzweil, *Singularity Is Near*, 143.

88 Ibid., 125, 126–127.

89 Ibid., 70.

90 Ibid., 144.

91 Ibid., 158, 197.

92 Kurzweil, "Featured Session."

93 Kurzweil, *Singularity Is Near*, 124–125.

94 Daisy Yuhas and Ferris Jabr, "Know Your Neurons: What Is the Ratio of Glia to Neurons in the Brain?," *Scientific American*, June 13, 2012, www.scientificamerican.com/blog /brainwaves/know-your-neurons-what-is-the-ratio-of-glia-to-neurons-in-the-brain/.

95 Anders Sandberg and Nick Bostrom, *Whole Brain Emulation: A Roadmap*, Technical Report no. 2008-3 (Oxford: Future of Humanity Institute, 2008), 80–81, www .fhi.ox.ac.uk/brain-emulation-roadmap-report.pdf. The same study suggests that Kurzweil's estimate of the human brain's memory capacity is far too low.

96 Hendricks, interview with author.

97 Rodney Brooks, "I, Rodney Brooks, Am a Robot," *IEEE Spectrum*, June 1, 2008, https://spectrum.ieee.org/i-rodney-brooks-am-a-robot.

98 Matthew Cobb, *The Idea of the Brain: The Past and Future of Neuroscience* (New York: Basic Books, 2020), 377–379.

99 Ibid., 4–5.

100 Ibid., 379–380.

101 Ibid., 381.

102 Ibid., 375–376.

103 Such arguments are based on the Church-Turing thesis, though going from the original thesis to the particular claim being made about AGI isn't always straightforward. See Massimo Pigliucci, "Mind Uploading: A Philosophical Counter-Analysis," in *Intelligence Unbound: The Future of Uploaded and Machine Minds*, eds. Russell Blackford and Damien Broderick (Chichester, UK: Wiley-Blackwell, 2014), 119–130; B. Jack Copeland, "The Church-Turing Thesis," *Stanford Encyclopedia of Philosophy* (Spring 2024), https://plato.stanford.edu/archives/spr2024/entries/church-turing/.

104 An analogy for the software engineers: even if the computers we have now were powerful enough, implementing a human-level AI on them might be like trying to write a website back end in Brainfuck or some other obfuscated language—technically possible, because it's Turing complete, but not a great way to approach the problem.

105 Chalmers, "The Singularity," 11.

106 Rodney Brooks, "Predictions Scorecard, 2024 January 01," RodneyBrooks.com (blog), January 1, 2024, https://rodneybrooks.com/predictions-scorecard-2024-january-01/.

107 Kurzweil, *Singularity Is Near*, 29.

108 W. Patrick McCray, *The Visioneers: How a Group of Elite Scientists Pursued Space Colonies, Nanotechnologies, and a Limitless Future* (Princeton, NJ: Princeton University Press, 2012), 77, 148–150.

109 Ibid., 171.

110 McCray, *Visioneers*, 187; J. Storrs Hall, *Where Is My Flying Car?* (San Francisco: Stripe Press, 2021), 59.

111 Hall, *Flying Car*, 60.

112 Ibid., 61, 304.

113 Ibid., 64.

114 Regis, "Meet the Extropians."

115 Kurzweil, *Singularity Is Near*, 226.

116 Ibid., 29.

117 Ibid., 352.

118 Ibid., 365.

119 Ibid., 375 (and many, many other places in the same book).

120 Some philosophers, like Chalmers, have even floated the possibility that under the right circumstances, the uploaded version of you wouldn't be a copy of you—it might just be you. (Chalmers, "The Singularity"; see also Pigliucci, "Mind Uploading," for an opposing view.)

121 Kurzweil, *Singularity Is Near*, 21.

122 Ibid., 364.

123 Ibid., 342–344.

124 Hendricks interview.

125 Kurzweil, *Singularity Is Near*, 365.

126 Ibid., 28.

127 K. Eric Drexler and Richard Smalley, "Nanotechnology: Drexler and Smalley Make the Case for and Against 'Molecular Assemblers,'" *Chemical & Engineering News* 81, no. 48 (2003): 37–42, https://doi.org/10.1021/cen-v081n048.p037.

128 Ibid.

129 Steven Edwards, *The Nanotech Pioneers: Where Are They Taking Us?* (Hoboken, NJ: Wiley, 2006), 201.

130 Drexler and Smalley, "Nanotechnology."

131 Stoddart, interview with author.

132 Ibid.

133 Kurzweil, *Singularity Is Near*, 240.

134 Ibid., 13.

135 Ibid., 241.

136 Ibid., 163, 341–342.

137 Kurzweil, "Featured Session."

138 Ibid.

139 Christophe Lécuyer, "Driving Semiconductor Innovation: Moore's Law at Fairchild and Intel," *Enterprise & Society* 23, no. 1 (2022): 133–163, https://doi.org/10.1017/eso.2020.38.

140 I have borrowed the phrase "rhetoric of inevitability" from the historian of science Mara Beller, who used it to describe the arguments deployed in the early days of quantum physics by Niels Bohr and others in favor of what later came to be known as the Copenhagen interpretation. For more on this, see Beller's excellent book *Quantum Dialogue: The Making of a Revolution* (Chicago: University of Chicago Press, 1999). For a less technical discussion of the same issues, see my own *What Is Real?* (New York: Basic Books, 2018).

141 Kurzweil, "Featured Session."

142 David Kushner, "How A.I. Could Reincarnate Your Dead Grandparents—or Wipe Out Your Kids," *Rolling Stone*, September 4, 2023, www.rollingstone.com/culture/culture-features/ai-bots-destroy-humanity-immortality-1234816682/.

143 Kurzweil, "Featured Session."

144 Ibid.

145 Kurzweil, *Singularity Is Near*, 32.

CHAPTER 3

1 Yudkowsky, "Pausing AI Developments."

2 Eliezer Yudkowsky, "AGI Ruin: A List of Lethalities," LessWrong, June 5, 2022, www.lesswrong.com/posts/uMQ3cqWDPHhjtiesc/agi-ruin-a-list-of-lethalities.

3 Peter Thiel, "The End of the Future," speech given at the Stanford Academic Freedom Conference, November 4–5, 2022, video, YouTube, November 18, 2022, www.youtube.com/watch?v=ibR_ULHYirs.

4 Sam Altman (@sama), Twitter (now X), February 3, 2023, https://twitter.com/sama/status/1621621724507938816.

5 Sam Altman (@sama), Twitter, archived October 18, 2023, at the Wayback Machine, https://web.archive.org/web/20231018095623/twitter.com/sama (lack of capitalization in the original).

6 Eliezer Yudkowsky, "There's No Fire Alarm for Artificial General Intelligence," Machine Intelligence Research Institute, October 13, 2017, https://intelligence.org/2017/10/13/fire-alarm/.

7 Many places, e.g., Lex Fridman, "#368—Eliezer Yudkowsky: Dangers of AI and the End of Human Civilization," March 30, 2023, in *Lex Fridman Podcast*, podcast, https://lexfridman.com/eliezer-yudkowsky/.

8 Yudkowsky interview.

9 Yudkowsky, "Pausing AI Developments."

10 Ibid.

11 From Eliezer Yudkowsky, "Fundamental Difficulties in Aligning Advanced AI," talk given at NYU, 2017, video, YouTube, September 1, 2017, www.youtube.com/watch?v=YicCAgjsky8.

12 First shows up in his 2003 paper: Nick Bostrom, "Ethical Issues in Advanced Artificial Intelligence," in *Cognitive, Emotive and Ethical Aspects of Decision Making in Humans and in Artificial Intelligence*, vol. 2, eds. I. Smit et al. (Tecumseh, Ontario: International Institute for Advanced Studies in Systems Research and Cybernetics, 2003), 12–17, https://nickbostrom.com/ethics/ai.

13 Rudyard Griffiths, "Nick Bostrom: 'I Don't Think the Artificial-Intelligence Train Will Slow Down,'" *Globe and Mail*, May 1, 2015, www.theglobeandmail.com/opinion/munk-debates/nick-bostrom-i-dont-think-the-artificial-intelligence-train-will-slow-down/article24222185/.

14 Nick Bostrom, *Superintelligence: Paths, Dangers, Strategies* (Oxford: Oxford University Press, 2017), 77 (for the quote), 94 (for the claim that a fast or moderate takeoff is more likely).

15 Griffiths, "Nick Bostrom."

16 Ibid.

17 Nick Bostrom, "The Superintelligent Will: Motivation and Instrumental Rationality in Advanced Artificial Agents," *Minds and Machines* 22, no. 2 (May 2012), https://nickbostrom.com/superintelligentwill.pdf.

18 Ibid. This is the "instrumental convergence thesis" described in section 2 of that paper. Yudkowsky endorses that thesis (along with the orthogonality thesis) in many places in his writing, e.g., Yudkowsky, "AGI Ruin."

19 Yudkowsky, "AGI Ruin."

20 Yudkowsky, "Fire Alarm."

21 Eliezer Yudkowsky, "Eliezer, the Person," August 31, 2000, archived February 5, 2001, at the Wayback Machine, https://web.archive.org/web/20010205221413/http://sysopmind.com/eliezer.html.

22 Eliezer Yudkowsky, "Bookshelf," 1999, archived February 5, 2001, at the Wayback Machine, https://web.archive.org/web/20010205060700/http://sysopmind.com/bookshelf.html.

23 Vernor Vinge, *True Names . . . and Other Dangers* (New York: Baen Books, 1987), 47.

24 Yudkowsky, "Eliezer, the Person" (emphasis his).

25 Eliezer Yudkowsky, "Staring into the Singularity 1.2.1," August 31, 2000, archived January 25, 2001, at the Wayback Machine, https://web.archive.org/web/20010125023900/http://sysopmind.com/singularity.html (emphasis his).

26 Extropians listserv archive, July 25–December 31, 1996, https://extropians.weidai.com/extropians.96/.

27 Eliezer Yudkowsky, "Re: Arrogent [*sic*] Bastards," Extropians listserv archive, December 14, 1996, https://extropians.weidai.com/extropians.96/4049.html.

28 Extropians listserv archive, July 25–December 31, 1996.

29 "Engines of Creation 2000 Confronting Singularity" (Foresight Institute, Palo Alto, CA, May 19–21, 2000), archived April 28, 2001, at the Wayback Machine, https://web.archive.org/web/20010428233716/https://foresight.org/SrAssoc/spring2000/.

30 "About the Institute," Singularity Institute for Artificial Intelligence, archived December 10, 2000, at the Wayback Machine, https://web.archive.org/web/20001210004500/http://singinst.org/about.html; Luke Muehlhauser, "AI Risk and Opportunity: Humanity's Efforts So Far," LessWrong, March 21, 2012, www.lesswrong.com/posts/i4susk4W3ieR5K92u/ai-risk-and-opportunity-humanity-s-efforts-so-far; "MIRI: Artificial Intelligence: The Danger of Good Intentions," Future of Life Institute, October 12, 2015, https://futureoflife.org/ai/ai-the-danger-of-good-intentions/; Yudkowsky interview.

31 Yudkowsky, "Eliezer, the Person" (emphasis his).

32 On Twitter, in February 2023, Yudkowsky said this was "late 2000 or so": Eliezer Yudkowsky (@ESYudkowsky), Twitter (now X), February 12, 2023, https://twitter.com/ESYudkowsky/status/1624551127873392641. This lines up with the timeline on the "Creating Friendly AI" paper, which was published in June 2001: Eliezer Yudkowsky, *Creating Friendly AI 1.0: The Analysis and Design of Benevolent Goal Architectures* (San Francisco: The Singularity Institute, 2001), https://intelligence.org/files/CFAI.pdf. Also see Future of Life Institute, "MIRI."

33 Declan Mccullagh, "Making HAL Your Pal," *Wired*, April 19, 2001, www.wired.com/2001/04/making-hal-your-pal/.

34 Eliezer S. Yudkowsky, "The AI-Box Experiment," Yudkowsky.net, 2002, accessed June 16, 2024, www.yudkowsky.net/singularity/aibox.

35 Eliezer Yudkowsky, "Shut Up and Do the Impossible!," LessWrong, October 8, 2008, www.lesswrong.com/posts/nCvvhFBaayaXyuBiD/shut-up-and-do-the-impossible.

36 See also "dril" (@dril), Twitter (now X), April 9, 2014, https://twitter.com/dril/status/454051061125640192.

37 Eliezer Yudkowsky, "The 'AI Box' Experiment," SL4 listserv archives, March 8, 2002, http://sl4.org/archive/0203/3132.html.

38 Yudkowsky, "Shut Up."

39 Machine Intelligence Research Institute Inc, Form 990-EZ for period ending December 2003, ProPublica Nonprofit Explorer, accessed June 16, 2024, https://projects.propublica.org/nonprofits/display_990/582565917/2004_06_EO%2F58-2565917_990EZ_200312.

40 George Packer, "No Death, No Taxes," *New Yorker*, November 20, 2011, www.newyorker.com/magazine/2011/11/28/no-death-no-taxes.

41 The Thiel Foundation, Form 990-PF for period ending December 2006, ProPublica Nonprofit Explorer, accessed June 16, 2024, https://projects.propublica.org/nonprofits/display_990/203846597/2007_06_PF%2F20-3846597_990PF_200612; "Top Contributors," Machine Intelligence Research Institute, accessed June 16, 2024, https://intelligence.org/topcontributors/.

42 "Singularity Summit," Machine Intelligence Research Institute, accessed June 16, 2024, https://intelligence.org/singularitysummit/.

43 Muehlhauser, "AI Risk and Opportunity."

44 "About," Overcoming Bias, accessed June 16, 2024, www.overcomingbias.com/about; Yudkowsky interview.

45 Eliezer Yudkowsky, preface to *Rationality: From AI to Zombies* (Berkeley, CA: Machine Intelligence Research Institute, 2015), www.readthesequences.com/#preface; see also Zach Weinersmith, "Bayesian," *Saturday Morning Breakfast Cereal*, November 8, 2020, www.smbc-comics.com/comic/bayesian.

46 Or perhaps even longer. This is just the word count of *Rationality: From AI to Zombies*, which is a collection of Yudkowsky's Sequences published from 2006 to 2009. There are at least a couple of posts he called "Sequences" that lie outside that time period.

47 "Bay Area Overcoming Bias Meetup," Meetup.com, archived February 14, 2009, at the Wayback Machine, https://web.archive.org/web/20090214184817/https://www.meetup.com/Bay-Area-Overcoming-Bias-Meetup/; Eliezer Yudkowsky, "Selecting Rationalist Groups," LessWrong, April 2, 2009, www.lesswrong.com/posts/ZEj9ATpv3P22LSmnC/selecting-rationalist-groups.

48 Overcoming Bias, "About"; Ruben Bloom, "A Brief History of LessWrong," LessWrong, May 31, 2019, www.lesswrong.com/posts/S69ogAGXcc9EQjpcZ/a-brief-history-of-lesswrong.

49 Archive of Eliezer Yudkowsky (@ESYudkowsky), Twitter, archived March 22, 2018, at the Wayback Machine, https://web.archive.org/web/20180322173754/https://twitter.com/ESYudkowsky.

50 Roko Mijic, "Solutions to the Altruist's Burden: The Quantum Billionaire Trick," LessWrong, July 23, 2010, https://basilisk.neocities.org/; RationalWiki, s.v. "Roko's Basilisk/Original post," last modified July 23, 2010, 21:03, https://rationalwiki.org/wiki/Roko%27s_basilisk/Original_post. Capitalization and formatting in the original.

51 Rob Bensinger, "A Few Misconceptions Surrounding Roko's Basilisk," LessWrong, October 5, 2015, www.lesswrong.com/posts/WBJZoeJypcNRmsdHx/a-few-misconceptions-surrounding-roko-s-basilisk.

52 Eliezer Yudkowsky, comment on "Roko's Basilisk," August 7, 2014, r/Futurology, Reddit, www.reddit.com/r/Futurology/comments/2cm2eg/comment/cjjbqqo/.

53 Yudkowsky interview.

54 Yudkowsky, *HPMOR*.

55 Ibid., ch. 6.

56 Eliezer Yudkowsky, "Above-Average AI Scientists," LessWrong, September 28, 2008, www.lesswrong.com/posts/9HGR5qatMGoz4GhKj/above-average-ai-scientists (ellipses in the original).

57 Eliezer Yudkowsky, "Normal Cryonics," LessWrong, January 19, 2010, www.lesswrong.com/posts/hiDkhLyN5S2MEjrSE/normal-cryonics.

58 Eliezer Yudkowsky, "Author's Note 119: Shameless Blegging," in *HPMOR*, March 10, 2015, https://hpmor.com/notes/119/.

59 Eliezer Yudkowsky, "Author's Notes: Ch. 87," in *HPMOR*, December 21, 2012, https://hpmor.com/notes/87/.

60 Luke Muehlhauser, "We Are Now the 'Machine Intelligence Research Institute' (MIRI)," Machine Intelligence Research Institute, January 30, 2013, https://intelligence.org/2013/01/30/we-are-now-the-machine-intelligence-research-institute-miri/.

61 Eliezer Yudkowsky, "What Do We Mean By 'Rationality'?," LessWrong, March 16, 2009, www.lesswrong.com/posts/RcZCwxFiZzE6X7nsv/what-do-we-mean-by-rationality-1.

62 Yudkowsky, "Preface."

63 Eliezer Yudkowsky, "Author's Notes, Ch. 78," in *HPMOR*, March 11, 2012, https://hpmor.com/notes/78/.

64 Anna Salamon, "Minicamps on Rationality and Awesomeness: May 11–13, June 22–24, and July 21–28," LessWrong, March 29, 2012, www.lesswrong.com/posts/fkhbBE2ZTSytvsy9x/minicamps-on-rationality-and-awesomeness-may-11-13-june-22.

65 Yudkowsky, "Author's Notes: Ch. 87."

66 Yudkowsky, "Staring into the Singularity."

67 "curiousepic," "Transcription of Eliezer's January 2010 Video Q&A," LessWrong, November 14, 2011, www.lesswrong.com/posts/YduZEfz8usGbJXN4x/transcription-of-eliezer-s-january-2010-video-q-and-a.

68 Will Henshall, "Time100 AI: Eliezer Yudkowsky," *Time*, September 7, 2023, https://time.com/collection/time100-ai/6309037/eliezer-yudkowsky/.

69 Thiel, "The End of the Future."

70 MIRI, "Top Contributors."

71 Ibid.; Jaan Tallinn, "List of Donations," Jaan.info, accessed September 1, 2024, https://jaan.info/philanthropy/donations.html.

72 "FTX Trading Ltd.: Case 22-11068," docket no. 13721, May 2, 2024, Kroll, https://restructuring.ra.kroll.com/FTX/Home-DocketInfo.

73 Barrat, *Our Final Invention*, 117.

74 Matthew Hutson, "Can We Stop Runaway A.I.?," *New Yorker*, May 16, 2023, www.newyorker.com/science/annals-of-artificial-intelligence/can-we-stop-the-singularity.

75 "The So-Called 'Godfather of the A.I.' Joins the Lead to Offer a Dire Warning About the Dangers of Artificial Intelligence," CNN, May 2, 2023, www.cnn.com/videos/tv/2023/05/02/the-lead-geoffrey-hinton.cnn.

76 Kevin Roose, "Inside the White-Hot Center of A.I. Doomerism," *New York Times*, July 11, 2023, www.nytimes.com/2023/07/11/technology/anthropic-ai-claude-chatbot.html; Krystal Hu, "Google Agrees to Invest up to $2 Billion in OpenAI Rival Anthropic," Reuters, October 27, 2023, www.reuters.com/technology/google-agrees-invest-up-2-bln-openai-rival-anthropic-wsj-2023-10-27/; Devin Coldewey, "Amazon Doubles Down on Anthropic, Completing Its Planned $4B Investment," *TechCrunch*, March 27, 2024, https://techcrunch.com/2024/03/27/amazon-doubles-down-on-anthropic-completing-its-planned-4b-investment/.

77 Gebru, interview with author.

78 Ibid.

79 Yann LeCun (@ylecun), Twitter (now X), April 23, 2023, https://twitter.com/ylecun/status/1650249953682370569.

80 Yann LeCun (@ylecun), Twitter (now X), April 23, 2023, https://twitter.com/ylecun/status/1650251352021377026; Yann LeCun (@ylecun), Twitter (now X), April 23, 2023, https://twitter.com/ylecun/status/1650288972956946433.

81 Jeremie Harris, "Oren Etzioni—The Case Against (Worrying About) Existential Risk from AI," June 16, 2021, in *Towards Data Science*, podcast, YouTube, www.youtube.com/watch?v=xtzxfjDDKfw.

82 Mitchell, interview with author.

83 Maciej Cegłowski, "Superintelligence: The Idea That Eats Smart People" (talk, Web Camp Zagreb, Croatia, October 29, 2016), https://idlewords.com/talks/superintelligence.htm.

84 This is also similar to one of the arguments given in Chalmers, "The Singularity," which Thorstad, "Belonging," responds to.

85 Tallinn, interview with author.

86 Cegłowski, "Superintelligence."

87 Ibid.

88 Ibid. Immediately after this, Cegłowski continues, "For all we know, human-level intelligence could be a tradeoff. Maybe any entity significantly smarter than a human being would be crippled by existential despair, or spend all its time in Buddha-like contemplation."

89 Melanie Mitchell, "Debates on the Nature of Artificial General Intelligence," *Science* 383, no. 6689 (March 21, 2024), https://doi.org/10.1126/science.ado7069.

90 Kevin Roose, "A Conversation with Bing's Chatbot Left Me Deeply Unsettled," *New York Times*, February 16, 2023, www.nytimes.com/2023/02/16/technology/bing-chatbot-microsoft-chatgpt.html.

91 Matt O'Brien, "Is Bing Too Belligerent? Microsoft Looks to Tame AI Chatbot," AP, February 16, 2023, https://apnews.com/article/technology-science-microsoft-corp-business-software-fb49e5d625bf37be0527e5173116bef3; Seth Lazar (@sethlazar), Twitter (now X), February 16, 2023, https://twitter.com/sethlazar/status/1626257535178280960.

92 To my great surprise, it appears that the first use of "hallucination" in this broad sense relating to neural networks arguably dates back to the mid-1980s. See Eric Daniel Mjolsness, "Neural Networks, Pattern Recognition, and Fingerprint Hallucination" (PhD diss., California Institute of Technology, 1986), http://doi.org/10.7907/M0VQ-DJ43. Thanks to John Wenz for finding this (and many other things).

93 Benjamin Weiser and Nate Schweber, "The ChatGPT Lawyer Explains Himself," *New York Times*, June 8, 2023, www.nytimes.com/2023/06/08/nyregion/lawyer-chatgpt-sanctions.html.

94 Ted Chiang, "ChatGPT Is a Blurry JPEG of the Web," *New Yorker*, February 9, 2023, www.newyorker.com/tech/annals-of-technology/chatgpt-is-a-blurry-jpeg-of-the-web.

95 Ibid.

96 Ilia Shumailov et al., "The Curse of Recursion: Training on Generated Data Makes Models Forget," arXiv:2305.17493 (2023), https://doi.org/10.48550/arXiv.2305.17493; Ross Anderson, "Will GPT Models Choke on Their Own Exhaust?," *Light Blue Touchpaper* (blog), June 6, 2023, www.lightbluetouchpaper.org/2023/06/06/will-gpt-models-choke-on-their-own-exhaust/.

97 Nick Bonyhady, "AI Companies Face 'Model Collapse.' They Should Pay to Fix It," *Australian Financial Review*, December 29, 2023, www.afr.com/technology/ai-companies-face-model-collapse-they-should-pay-to-fix-it-20231228-p5eu0r.

98 Colin Fraser, "ChatGPT: Automatic Expensive BS at Scale," Medium, January 27, 2023, https://medium.com/@colin.fraser/chatgpt-automatic-expensive-bs-at-scale-a113692b13d5.

99 John Seabrook, "The Next Word," *New Yorker*, October 14, 2019, www.newyorker.com/magazine/2019/10/14/can-a-machine-learn-to-write-for-the-new-yorker (emphasis in original).

100 Dwarkesh Patel, "Ilya Sutskever (OpenAI Chief Scientist)—Building AGI, Alignment, Spies, Microsoft, & Enlightenment," in *Dwarkesh Podcast*, YouTube, March 27, 2023, www.youtube.com/watch?v=Yf1o0TQzry8.

101 Karen Hao and Charlie Warzel, "Inside the Chaos at OpenAI," *The Atlantic*, November 19, 2023, www.theatlantic.com/technology/archive/2023/11/sam-altman-open-ai-chatgpt-chaos/676050/; Cade Metz, "OpenAI's Chief Scientist and Co-Founder Is Leaving the Company," *New York Times*, May 14, 2024, www.nytimes.com/2024/05/14/technology/ilya-sutskever-leaving-openai.html.

102 Michaël Trazzi, "Ethan Caballero on Why Scale Is All You Need," May 5, 2022, on *The Inside View*, podcast, https://theinsideview.ai/ethan.

103 Fraser, "ChatGPT." See also this fascinating empirical study, which strongly suggests that scale can't be all you need: Vishaal Udandarao et al., "No 'Zero-Shot' Without Exponential Data: Pretraining Concept Frequency Determines Multimodal Model Performance," arXiv:2404.04125 (2024), https://doi.org/10.48550/arXiv.2404.04125.

104 Shana Lynch, "Is GPT-3 Intelligent? A Directors' Conversation with Oren Etzioni," Stanford University Human-Centered Artificial Intelligence, October 1, 2020, https://hai.stanford.edu/news/gpt-3-intelligent-directors-conversation-oren-etzioni.

105 Wilfred Chan, "Researcher Meredith Whittaker Says AI's Biggest Risk Isn't 'Consciousness'—It's the Corporations That Control Them," *Fast Company*, May 5, 2023, www.fastcompany.com/90892235/researcher-meredith-whittaker-says-ais-biggest-risk-isnt-consciousness-its-the-corporations-that-control-them.

106 Will Douglas Heaven, "How Existential Risk Became the Biggest Meme in AI," *MIT Technology Review*, June 19, 2023, www.technologyreview.com/2023/06/19/1075140/how-existential-risk-became-biggest-meme-in-ai/.

107 Ibid.

108 Chan, "Researcher Meredith Whittaker."

109 Julia Angwin et al., "Machine Bias," ProPublica, May 23, 2016, www.propublica.org/article/machine-bias-risk-assessments-in-criminal-sentencing; Julia Angwin et al., "What Algorithmic Injustice Looks Like in Real Life," ProPublica, May 25, 2016, www.propublica.org/article/what-algorithmic-injustice-looks-like-in-real-life.

110 Ibid.

111 "Diversity in High Tech," US Equal Employment Opportunity Commission, May 18, 2016, www.eeoc.gov/special-report/diversity-high-tech; Ashton Jackson, "Black Employees Make Up Just 7.4% of the Tech Workforce—These Nonprofits Are Working to Change That," CNBC, February 24, 2022, www.cnbc.com/2022/02/24/jobs-for-the-future-report-highlights-need-for-tech-opportunities-for-black-talent.html.

112 Khari Johnson, "AI Ethics Pioneer's Exit from Google Involved Research into Risks and Inequality in Large Language Models," *VentureBeat*, December 3, 2020, https://venturebeat.com/2020/12/03/ai-ethics-pioneers-exit-from-google-involved-research-into-risks-and-inequality-in-large-language-models/; Karen Hao, "We Read the Paper That Forced Timnit Gebru Out of Google. Here's What It Says," *MIT Technology Review*, December 4, 2020, www.technologyreview.com/2020/12/04/1013294/google-ai-ethics-research-paper-forced-out-timnit-gebru/.

113 Krystal Hu, "ChatGPT Sets Record for Fastest-Growing User Base," Reuters, February 2, 2023, www.reuters.com/technology/chatgpt-sets-record-fastest-growing-user-base-analyst-note-2023-02-01/.

114 Hao, "We Read the Paper."

115 Johnson, "AI Ethics Pioneer's Exit"; Hao, "We Read the Paper."

116 Megan Rose Dickey, "Google Fires Top AI Ethics Researcher Margaret Mitchell," *TechCrunch*, February 19, 2021, https://techcrunch.com/2021/02/19/google-fires-top-ai-ethics-researcher-margaret-mitchell/; Nico Grant, Dina Bass, and Josh Eidelson, "Google Fires Researcher Meg Mitchell, Escalating AI Saga," Bloomberg, February 19, 2021, www.bloomberg.com/news/articles/2021-02-19/google-fires-researcher-meg-mitchell-escalating-ai-saga.

117 Kyle Wiggers, "Google Trained a Trillion-Parameter AI Language Model," *VentureBeat*, January 12, 2021, https://venturebeat.com/2021/01/12/google-trained-a-trillion-parameter-ai-language-model/; William Fedus, Barret Zoph, and Noam Shazeer, "Switch Transformers: Scaling to Trillion Parameter Models with Simple and Efficient Sparsity," arXiv:2101.03961 (2021), https://doi.org/10.48550/arXiv.2101.03961.

118 Thaddeus L. Johnson and Natasha N. Johnson, "Police Facial Recognition Technology Can't Tell Black People Apart," *Scientific American*, May 18, 2023, www.scientificamerican.com/article/police-facial-recognition-technology-cant-tell-black-people-apart/.

119 Dhruv Mehrotra et al., "How We Determined Crime Prediction Software Disproportionately Targeted Low-Income, Black, and Latino Neighborhoods," *The Markup*, December 2, 2021, https://themarkup.org/show-your-work/2021/12/02/how-we-determined-crime-prediction-software-disproportionately-targeted-low-income-black-and-latino-neighborhoods.

120 Kaveh Waddell, "How Algorithms Can Bring Down Minorities' Credit Scores," *The Atlantic*, December 2, 2016, www.theatlantic.com/technology/archive/2016/12/how-algorithms-can-bring-down-minorities-credit-scores/509333/; Will Douglas Heaven, "Bias Isn't the Only Problem with Credit Scores—and No, AI Can't Help," *MIT Technology Review*, June 17, 2021, www.technologyreview.com/2021/06/17/1026519/racial-bias-noisy-data-credit-scores-mortgage-loans-fairness-machine-learning/.

121 CNN, "The So-Called 'Godfather of the A.I.'"

122 Eliezer Yudkowsky (@ESYudkowsky), Twitter (now X), October 24, 2022, https://twitter.com/ESYudkowsky/status/1584719334873894913.

123 Yudkowsky interview.

124 Chan, "Researcher Meredith Whittaker."

125 Harris, "Oren Etzioni."

126 François Chollet (@fchollet), Twitter (now X), August 24, 2022, https://twitter.com/fchollet/status/1562506185873702916.

127 Gebru interview.

128 As quoted in Gorman Beauchamp, "The Frankenstein Complex and Asimov's Robots," *Mosaic* 13, no. 3/4 (1980): 83–94, www.jstor.org/stable/24780264.

129 The story about the escaped robot is "Little Lost Robot," first published in 1947 and later collected in Asimov, *I, Robot*; the story about the dreaming robot is "Robot Dreams," first published in 1986.

130 David Golumbia, "The Great White Robot God," Medium, January 21, 2019, https://davidgolumbia.medium.com/the-great-white-robot-god-bea8e23943da.

131 Stephen Jay Gould, "Curveball," *New Yorker*, November 28, 1994, https://chance.dartmouth.edu/course/topics/curveball.html.

132 Mark Dery, *Escape Velocity: Cyberculture at the End of the Century* (New York: Grove Press, 1996), 307.

133 Anders Hove, "Extropians Take Their Cue from Bigotry," *The Tech*, August 20, 1997, archived March 1, 2023, at the Wayback Machine, https://web.archive.org/web/20230301090204/http://tech.mit.edu/V117/N30/anders.30c.html (via Thorstad, "Belonging"); Anna Dirks, "MIT Extropians Anger Many," *MIT Observer*, September 3, 1997, https://web.mit.edu/observer/www/1-1/articles/ad1.html.

134 Mike Lorrey, "Re: Are Extropians Libertarian? (Was RE: Learn to Shoot)," Extropians listserv archive, October 29, 2000, https://diyhpl.us/~bryan/irc/extropians/extracted-extropians-archive/archive/0010/70932.html (via Thorstad, "Belonging").

135 Goertzel, "Extropian Creed."

136 John K. Clark, "The Extropian Principles," Extropians listserv archive, July 29, 1996, www.lucifer.com/exi-lists/extropians.96/0064.html.

137 Nick Bostrom, "Re: Offending People's Minds," Extropians listserv archive, August 24, 1996, www.lucifer.com/exi-lists/extropians.96/0441.html (racial slur unredacted in original).

138 Nick Bostrom, "Apology for an Old Email," January 9, 2023, https://nickbostrom.com/oldemail.pdf.

139 Deb Raji (@rajiinio), Twitter (now X), January 11, 2023, https://twitter.com/rajiinio/status/1613341900043612160.

140 Bostrom, "Re: Offending People's Minds."

141 Thorstad, "Belonging" (emphasis in original).

142 Mencius [Curtis Yarvin], comment on "Nationalist Moral Chauvinism," by Will Wilkinson, WillWilkinson.net, February 11, 2008, archived August 12, 2013, at the Wayback Machine, https://web.archive.org/web/20130812042830/http://willwilkinson.net/flybottle/2008/02/09/nationalist-moral-chauvinism/. NB: The genetic superiority of the minds of Ashkenazi Jews is an article of faith for adherents of modern forms of "scientific" racism like "human biodiversity." As an Ashkenazi Jew myself, I can tell you this is certainly false.

143 Robin Hanson, "Reply to Moldbug," Overcoming Bias, May 22, 2009, www.overcomingbias.com/p/reply-to-moldbughtml; Scott Alexander [Scott Siskind], "The Anti-Reactionary FAQ," *Slate Star Codex* (blog), October 20, 2013, https://slatestarcodex.com/2013/10/20/the-anti-reactionary-faq/.

144 Eliezer Yudkowsky (Optimize Literally Everything), "This isn't going to work, but for the record . . .," Tumblr, April 8, 2016, https://yudkowsky.tumblr.com/post/142497361345/this-isnt-going-to-work-but-for-the-record-and.

145 Mencius Moldbug [Curtis Yarvin], "Divine-Right Monarchy for the Modern Secular Intellectual," *Unqualified Reservations* (blog), March 19, 2010, www.unqualified-reservations.org/2010/03/divine-right-monarchy-for-modern/.

146 Mencius Moldbug [Curtis Yarvin], "Why I Am Not a White Nationalist," *Unqualified Reservations* (blog), November 22, 2007, www.unqualified-reservations.org/2007/11/why-i-am-not-white-nationalist/.

147 Mencius Moldbug [Curtis Yarvin], "Charles Stross Discovers the Cathedral," *Unqualified Reservations* (blog), February 8, 2013, www.unqualified-reservations.org/2013/02/charles-stross-discovers-cathedral/.

148 Peter Thiel, "The Education of a Libertarian," *Cato Unbound*, April 13, 2009, www.cato-unbound.org/2009/04/13/peter-thiel/education-libertarian/.

149 Peter Thiel, "Your Suffrage Isn't in Danger. Your Other Rights Are," *Cato Unbound*, May 1, 2009, www.cato-unbound.org/2009/05/01/peter-thiel/suffrage-isnt-danger-other-rights-are/.

150 Barton Gellman, "Peter Thiel Is Taking a Break from Democracy," *The Atlantic*, November 9, 2023, www.theatlantic.com/politics/archive/2023/11/peter-thiel-2024-election-politics-investing-life-views/675946/.

151 Colin Lecher, "Alt-right Darling Mencius Moldbug Wanted to Destroy Democracy. Now He Wants to Sell You Web Services," *The Verge*, February 21, 2017, www.theverge.com/2017/2/21/14671978/alt-right-mencius-moldbug-urbit-curtis-yarvin-tlon; Joseph Bernstein, "Here's How Breitbart and Milo Smuggled White Nationalism into the Mainstream," *BuzzFeed News*, October 5, 2017, www.buzzfeednews.com/article/josephbernstein/heres-how-breitbart-and-milo-smuggled-white-nationalism; Corey Pein, "Mouthbreathing Machiavellis Dream of a Silicon Reich," *The Baffler*, May 19, 2014, https://thebaffler.com/latest/mouthbreathing-machiavellis.

152 Bernstein, "Here's How Breitbart."

153 James Pogue, "Inside the New Right, Where Peter Thiel Is Placing His Biggest Bets," *Vanity Fair*, April 20, 2022, www.vanityfair.com/news/2022/04/inside-the-new-right-where-peter-thiel-is-placing-his-biggest-bets; Andrew Prokop, "Curtis Yarvin Wants American Democracy Toppled. He Has Some Prominent Republican Fans," *Vox*, October 24, 2022, www.vox.com/policy-and-politics/23373795/curtis-yarvin-neoreaction-redpill-moldbug.

154 Saini, interview with author.

155 "There's absolutely no biological difference between races. There's no genetic support for the idea that different races of humans are different in any really meaningful way," says Daniel Jordan, the director of computational genomics at the Mount Sinai Center for Genomic Data Analytics. Modern humans have only been around for about one hundred thousand years, which isn't very long at all in evolutionary terms. We are all descended from a small population originating in Africa. That population split and spread across the globe, but there has always been contact between neighboring populations—and sometimes distant populations—ensuring a great deal of genetic intermingling over the course of human history. "As a result of the relatively recent common origin of modern humans and the repeated mixing of groups, the alleles [different variants of the same gene] carried by people living all over the globe show little differentiation," wrote a team of scientists in a 2023 National Academies report on population descriptors in genetics.

The most information that you can read off of these small variations in genetics, says Jordan, is something about the migration history of a person's distant ancestors over the past hundred millennia, and that information isn't tied to deep biological differences. "There have been a few things that looked like they might point to differences in fundamental biology between continental ancestry groups, and they have all turned out to not really be real," he says. "There's really no reliable evidence that there's anything biologically meaningful about these groupings other than the shared migration history." Even these groupings are somewhat fuzzy, says the National Academies report, and they don't mean what we might think they do. "Human allele frequencies tend to vary continuously with geographic distance (isolation by distance), with slightly larger differences seen across long-term inhibitors of migration such as oceans or mountains. These geographic boundaries do not correspond to racial groupings."

Race isn't a biological reality; it's a social construct. "Our social conceptions of race and ethnicity do not match the underlying biological and genetic variation within our species, and we should never confuse the things that were created for the purposes of oppressing people with the nature of that biological and genetic variation," says evolutionary biologist Joseph Graves. Humans are extremely genetically similar to one another, even when compared to our nearest relatives. "There is more genetic variation within one tribe of wild chimpanzees than has been observed within all existing humans," Graves writes. "The majority of genetic variation in humans occurs between individuals, without regard to membership in a socially constructed race." In other words, there's greater genetic variety within races than there is between them.

Nor is there any indication that the genetic traits associated with the social construct of race—e.g., skin tone—have any biological connection with the genetic markers that are associated with performance on IQ tests. Instead, the opposite is true. "The genetic variation related to IQ is broadly distributed across the genome, rather than being clustered around a few spots, as is the nature of the variation responsible for skin pigmentation," write the geneticists Ewan Birney, Jennifer Raff, Adam Rutherford, and Aylwyn Scally. "These very different patterns for these two traits mean that the genes responsible for determining skin pigmentation cannot be meaningfully associated with the genes currently known to be linked to IQ. . . . Any apparent population differences in IQ scores are more easily explained by cultural and environmental factors than they are by genetics."

Despite the lack of evidence for their claims, it's understandable that racists would latch onto genetics as a way to find scientific justification for their views—there's a long history of doing so. "The field of human genetics was fundamentally founded as an offshoot of eugenics," says Jordan. "A lot of the equations and concepts that form the basis for our understanding of human population genetics were first applied to humans explicitly with eugenics in mind." This goes back to the historical roots of the idea of race as a biological concept in the nineteenth century. "The rise of racial ideology coincided with the rise of Darwinism (specifically, a misunderstanding of how Darwin's observations applied to humans) and the development of social institutions that exploited human biological differences for profit," wrote Graves. "This meant that a person's West African ancestry could be used as the sole reason to reduce him to chattel slavery, and that a group's American Indian ancestry in itself provided sufficient reason for the partial extermination of their population and seizure of their land." Race as a coherent biological category is simply a lie, one manufactured for a particular set of ends. "The biological concept of race in humans was created to support settler colonialism and slavery, and has always been entangled with racist institutions, policies, and practices," write the authors of the National Academies report. "The use of race as a population descriptor in scientific research therefore has caused incalculable confusion and harm."

References: Ewan Birney et al., "Race, Genetics and Pseudoscience: An Explainer," *Ewan's Blog: Bioinformatician at Large*, October 24, 2019, https://ewanbirney.com

/2019/10/race-genetics-and-pseudoscience-an-explainer.html; Joseph L. Graves, "The Biological Case Against Race," *American Outlook* (Spring 2002), archived November 15, 2006, at the Wayback Machine, https://web.archive.org/web/20061115040049/http://www.heartland.org/pdf/12721n.pdf; Jordan, interview with author; National Academies of Sciences, Engineering, and Medicine, *Using Population Descriptors in Genetics and Genomics Research: A New Framework for an Evolving Field* (Washington, DC: National Academies Press, 2023), https://doi.org/10.17226/26902.

156 Scott Siskind, "ACX Survey Results 2024," *Astral Codex Ten* (blog), April 19, 2024, www.astralcodexten.com/p/acx-survey-results-2024.

157 To the best of my knowledge, Siskind has not publicly confirmed—or denied—that the email Brennan posted was genuine.

158 Topher Brennan (@TopherTBrennan), Twitter (now X), February 17, 2021, archived February 17, 2021, at the Wayback Machine, https://web.archive.org/web/20210217195335/https://twitter.com/TopherTBrennan/status/1362108632070905857; screenshots posted to Imgur at: https://imgur.com/a/gWeIK6c; "Aster" (@ArsonAtDennys), Twitter (now X), February 17, 2021, https://twitter.com/ArsonAtDennys/status/1362153191102677001; Emil O. W. Kirkegaard, "Backstabber Brennan Knifes Scott Alexander with 2014 Email," *Clear Language, Clear Mind* (blog), February 17, 2021, https://emilkirkegaard.dk/en/2021/02/backstabber-brennan-knifes-scott-alexander-with-2014-email/.

159 Scott Siskind, "How Should We Think About Race and 'Lived Experience'?," *Astral Codex Ten* (blog), March 7, 2024, www.astralcodexten.com/p/how-should-we-think-about-race-and.

160 Scott Siskind, "Galton, Ehrlich, Buck," *Astral Codex Ten* (blog), May 15, 2023, www.astralcodexten.com/p/galton-ehrlich-buck. Scott Alexander [Scott Siskind], "The Atomic Bomb Considered as Hungarian High School Science Fair Project," *Slate Star Codex* (blog), May 26, 2017, https://slatestarcodex.com/2017/05/26/the-atomic-bomb-considered-as-hungarian-high-school-science-fair-project/.

161 Gebru interview.

162 Jacqueline Bryk (@RuffleJax), Twitter (now X), June 19, 2018, https://twitter.com/RuffleJax/status/1009140252085243906.

163 Jessica Taylor, "My Experience at and Around MIRI and CFAR (Inspired by Zoe Curzi's Writeup of Experiences at Leverage)," LessWrong, October 16, 2021, www.lesswrong.com/posts/MnFqyPLqbiKL8nSR7/my-experience-at-and-around-miri-and-cfar-inspired-by-zoe; Jessica Taylor (jessicata), user profile, LessWrong, accessed June 17, 2024, www.lesswrong.com/users/jessica-liu-taylor?from=post_header.

164 Taylor, "My Experience."

165 Qiaochu Yuan (@QiaochuYuan), "sometimes it's nice to put on a movie . . .," Thread Reader, July 1, 2022, https://threadreaderapp.com/thread/1542755798169686017.

166 Gebru interview.

167 Luke Darby, "Private Jets, Parties and Eugenics: Jeffrey Epstein's Bizarre World of Scientists," *The Guardian*, August 19, 2019, www.theguardian.com/us-news/2019/aug/18/private-jets-parties-and-eugenics-jeffrey-epsteins-bizarre-world-of-scientists.

168 Janet Maslin, "A Bizarre Tale of the Rise and Fall of an Elitist Sperm Bank," *New York Times*, June 2, 2005, www.nytimes.com/2005/06/02/books/a-bizarre-tale-of-the-rise-and-fall-of-an-elitist-sperm-bank.html.

169 James B. Stewart, Matthew Goldstein, and Jessica Silver-Greenberg, "Jeffrey Epstein Hoped to Seed Human Race with His DNA," *New York Times*, July 31, 2019, www.nytimes.com/2019/07/31/business/jeffrey-epstein-eugenics.html.

170 Peter Aldhous, "Jeffrey Epstein's Links to Scientists Are Even More Extensive Than We Thought," *BuzzFeed News*, August 26, 2019, www.buzzfeednews.com/article/peteraldhous/jeffrey-epstein-science-donations-apologies-statements.

171 "A Timeline of the Jeffrey Epstein, Ghislaine Maxwell Scandal," AP, June 28, 2022, https://apnews.com/article/epstein-maxwell-timeline-b9f15710fabb72e8581c71e94acf513e.

172 "Wynne," "An Open Letter to Vitalik Buterin," Medium, June 14, 2021, https://fredwynne.medium.com/an-open-letter-to-vitalik-buterin-ce4681a7dbe; Ellen Huet, "The Real-Life Consequences of Silicon Valley's AI Obsession," Bloomberg, March 7, 2023, www.bloomberg.com/news/features/2023-03-07/effective-altruism-s-problems-go-beyond-sam-bankman-fried; Charlotte Alter, "Effective Altruism Promises to Do Good Better. These Women Say It Has a Toxic Culture of Sexual Harassment and Abuse," *Time*, February 3, 2023, https://time.com/6252617/effective-altruism-sexual-harassment/; Kelsey Piper, "Why Effective Altruism Struggles on Sexual Misconduct," *Vox*, February 15, 2023, www.vox.com/future-perfect/2023/2/15/23601143/effective-altruism-sexual-harassment-misconduct; Shekinah Alegra, "Why I Define My Experience at the Monastic Academy as Sexual Assault," LessWrong, May 24, 2022, www.lesswrong.com/posts/zbQapwn2hGNozcGsD/why-i-define-my-experience-at-the-monastic-academy-as-sexual.

173 Sonia Joseph (@soniajoseph_), Twitter (now X), November 13, 2022, https://twitter.com/soniajoseph_/status/1591601317306593280.

174 Huet, "Real-Life Consequences."

175 Keerthana Gopalakrishnan (@keerthanpg), Twitter (now X), November 12, 2022, https://twitter.com/keerthanpg/status/1591515891359186944.

176 Scott Siskind (slatestarscratchpad), "[content warning: sexual harassment, suicide]. . .," Tumblr, June 22, 2018, www.tumblr.com/slatestarscratchpad/175157697076/content-warning-sexual-harassment-suicide-i.

177 Scott Siskind, "I Do Not Understand 'Rape Culture,'" *Slate Star Codex* (blog), April 19, 2013, https://slatestarcodex.com/2013/04/19/i-do-not-understand-rape-culture/.

178 Reich, interview with author.

179 John Horgan, "AI Visionary Eliezer Yudkowsky on the Singularity, Bayesian Brains and Closet Goblins," *Scientific American*, March 1, 2016, www.scientificamerican.com/blog/cross-check/ai-visionary-eliezer-yudkowsky-on-the-singularity-bayesian-brains-and-closet-goblins/.

180 Yuan (@QiaochuYuan), "sometimes it's nice to put on a movie . . .," (ellipses in original).

181 Eliezer Yudkowsky (@ESYudkowsky), Twitter (now X), March 6, 2023, https://twitter.com/ESYudkowsky/status/1632726781785690113; Eliezer Yudkowsky (@ESYudkowsky), Twitter (now X), March 14, 2023, https://twitter.com/ESYudkowsky/status/1635577836525469697; Kevin Roose, "Why an Octopus-Like Creature Has Come to Symbolize the State of A.I.," *New York Times*, May 30, 2023, www.nytimes.com/2023/05/30/technology/shoggoth-meme-ai.html; "AI Notkilleveryoneism Memes" (@AISafetyMemes), X profile, accessed June 17, 2024, https://twitter.com/AISafetyMemes; Scott Siskind, "Janus' Simulators," *Astral Codex Ten* (blog), January 26, 2023, www.astralcodexten.com/p/janus-simulators.

182 Eliezer Yudkowsky, personal communication.

183 Fridman, "#368—Eliezer Yudkowsky."

184 "curiousepic," "Transcription."

185 Tom Chivers, *The AI Does Not Hate You: Superintelligence, Rationality and the Race to Save the World* (London: Weidenfeld & Nicolson, 2019), 227.

186 Gebru interview.

CHAPTER 4

1 Sandberg, interview with author.

2 Yudkowsky, "Normal Cryonics" (emphasis his).

3 Kate Golembiewski, "Life After Death? Cryonicists Try to Defy Mortality by Freezing Bodies," *Discover*, October 14, 2022, www.discovermagazine.com/technology/will-cryonically-frozen-bodies-ever-be-brought-back-to-life.

4 Hendricks interview.

5 Sandberg interview.

6 Hendricks interview.

7 Ord, interview with author.

8 Ibid.

9 Ord on climate change: David Thorstad, "Exaggerating the Risks (Part 2: Ord on Climate Risk)," *Reflective Altruism* (blog), December 1, 2022, https://ineffectivealtruismblog.com/2022/12/01/exaggerating-the-risks-part-2-ord-on-climate-risk/; Ord on nuclear war: Robert Wiblin, "A Framework for Comparing Global Problems in Terms of Expected Impact," 80,000 Hours, April 2016, updated October 2019, https://80000hours.org/problem. This is also all in Toby Ord, *The Precipice: Existential Risk and the Future of Humanity* (New York: Hachette Books, 2020).

10 Ord interview.

11 MacAskill, *What We Owe*, 136.

12 Ibid., 137.

13 Robert Wiblin, Arden Koehler, and Keiran Harris, "Toby Ord on the Precipice and Humanity's Potential Futures," 80,000 Hours, March 7, 2020, https://80000hours.org/podcast/episodes/toby-ord-the-precipice-existential-risk-future-humanity/#climate-change-005108. NB: There's a full transcript on that page, and I took this from the transcript.

14 Émile P. Torres, "What 'Longtermism' Gets Wrong About Climate Change," *Bulletin of the Atomic Scientists*, November 22, 2022, https://thebulletin.org/2022/11/what-longtermism-gets-wrong-about-climate-change/.

15 Andrew Watson, private email correspondence, January 25, 2024.

16 See Eric Schliesser, "On What We Owe the Future [No, Not on SBF/FTX!], Part 1," *Digressions&Impressions* (blog), November 24, 2022, https://digressionsnimpressions.typepad.com/digressionsimpressions/2022/11/on-what-we-owe-the-future-no-not-on-sbfftx.html.

17 Watson email.

18 PETM was fifty-six million years ago, and the global mean temperature was about 27.2–34.5°C, i.e., about 13–21°C higher than the preindustrial average. Gordon N. Inglis et al., "Global Mean Surface Temperature and Climate Sensitivity of the Early Eocene Climatic Optimum (EECO), Paleocene–Eocene Thermal Maximum (PETM), and Latest Paleocene," *Climate of the Past* 16, no. 5 (October 26, 2020), https://cp.copernicus.org/articles/16/1953/2020/. Another article saying the PETM average was about 35°C: A. Sluijs et al., "Warming, Euxinia and Sea Level Rise During the Paleocene–Eocene Thermal Maximum on the Gulf Coastal Plain: Implications for Ocean Oxygenation and Nutrient Cycling," *Climate of the Past* 10, no. 4 (July 25, 2014), https://cp.copernicus.org/articles/10/1421/2014/cp-10-1421-2014.html. Other estimates put the warming closer to 15°C. See diagram here: (Text says 5–8°C, but that's clearly an error. That's referring to the rise relative to the background *at the time* of the PETM, not now.) Michon Scott and Rebecca Lindsey, "What's the Hottest Earth's Ever Been?," Climate.gov, November 22, 2023, www.climate.gov/news-features/climate-qa/whats-hottest-earths-ever-been. This also puts PETM warming at about 15°C, and also says that 20°C might be as much as five hundred million years ago: Paul Voosen, "A 500-Million-Year Survey of Earth's Climate Reveals Dire Warning for Humanity," *Science*, May 22, 2019, www.science.org/content/article/500-million-year-survey-earths-climate-reveals-dire-warning-humanity. No ice caps during PETM, palm trees and crocodiles in the Arctic Circle: Scott and Lindsey, ""What's the Hottest Earth's Ever Been?" The Arctic Ocean was 76°F during PETM, swamp forests in Canadian high arctic, giant tortoises, flying lemurs: Peter Brannen, *The Ends of the World: Volcanic Apocalypses, Lethal Oceans, and Our Quest to Understand Earth's Past Mass Extinctions* (New York: Ecco, 2017), 221. Equatorial water temps at PETM were basically body temp, 36°C; waters off the coast of Antarctica were 20°C: Timothy Bralower and David Bice, "Ancient Climate Events: Paleocene Eocene Thermal Maximum," in "Earth 103: Earth in the Future," Penn State and NASA, accessed January 25, 2024, www.e-education.psu.edu/earth103/node/639. Sea surface temps higher than 36°C at the tropics: Joost Frieling et al., "Extreme Warmth and Heat-Stressed Plankton in the Tropics During the Paleocene-Eocene Thermal Maximum," *Science*

Advances 3, no. 3 (March 3, 2017), www.science.org/doi/10.1126/sciadv.1600891. Sea level 50–200 meters higher than present at PETM: App Sluijs et al., "Eustatic Variations During the Paleocene-Eocene Greenhouse World," *Paleooceanography and Paleoclimatology* 23, no. 4 (December 19, 2008), https://agupubs.onlinelibrary.wiley.com/doi/full/10.1029/2008PA001615. Average summer air temperatures on land around human body temperature, well outside the tropics: Jessica E. Tierney et al., "Spatial Patterns of Climate Change Across the Paleocene–Eocene Thermal Maximum," *Proceedings of the National Academy of Sciences* 119, no. 42 (October 18, 2022), www.pnas.org/doi/10.1073/pnas.2205326119.

19 Tim Stephens, "High-Fidelity Record of Earth's Climate History Puts Current Changes in Context," UC Santa Cruz, news release, September 10, 2020, https://news.ucsc.edu/2020/09/climate-variability.html.

20 Watson email.

21 Reader, I ate one.

22 See pages 1 and 9 of CEA's 990: Centre for Effective Altruism USA Inc., return of organization exempt from income tax (Form 990, 2021), accessed January 25, 2024, https://ev.org/wp-content/uploads/2023/11/FINAL-2022-Form-990-signed-CEA.pdf. I found that link here: "EV US Annual Reports," Effective Ventures, accessed January 25, 2024, https://ev.org/evf-usa-annual-reports/ (it's listed as their 2022 990 there). Note that the line between CEA and Effective Ventures is blurry; the website says EV but the 990 says CEA.

23 There were at least four major grants from Open Philanthropy to CEA that fiscal year, for a total of about $60 million, definitely most of CEA's operating budget that fiscal year: "Effective Ventures Foundation—Event Venue," Open Philanthropy, December 2021, www.openphilanthropy.org/grants/effective-ventures-foundation-uk-event-venue/. For more on the abbey, Wikipedia, s.v. "Wytham Abbey," last modified May 12, 2024, 21:53, https://en.wikipedia.org/wiki/Wytham_Abbey; Wytham Abbey, www.wythamabbey.org. "Effective Altruism Funds—Regranting Support," Open Philanthropy, February 2022, www.openphilanthropy.org/grants/effective-altruism-funds-re-granting-support/; "Centre for Effective Altruism—General Support (2022)," Open Philanthropy, March 2022, www.openphilanthropy.org/grants/centre-for-effective-altruism-general-support-2022/; "Centre for the Governance of AI—AI Field Building," Open Philanthropy, December 2021, www.openphilanthropy.org/grants/centre-for-the-governance-of-ai-ai-field-building/. There were a few smaller grants, too, which sum up to about $1.5 million, all visible here: "Centre for Effective Altruism," Grants, Open Philanthropy, accessed January 25, 2024, www.openphilanthropy.org/grants/?q=centre+for+effective+altruism; "Effective Ventures," Grants, Open Philanthropy, accessed January 25, 2024, www.openphilanthropy.org/grants/?q=effective+ventures.

24 Jeroen Willems, "Why Did CEA Buy Wytham Abbey?," Effective Altruism Forum, December 6, 2022, https://forum.effectivealtruism.org/posts/xof7iFB3uh8Kc53bG/why-did-cea-buy-wytham-abbey; "nikos," "Reflections on Wytham Abbey," Effective Altruism Forum, January 10, 2023, https://forum.effectivealtruism.org/posts/76dQ6YfBuLzJDdTgz/reflections-on-wytham-abbey.

25 Rupert Neate, "Wytham Abbey Put Up for Sale for £15m by Effective Altruism Group EVF," *The Guardian*, May 12, 2024, www.theguardian.com/business/article/2024/may/12/wytham-abbey-sale-effective-altruism-group-evf.

26 Kulish, "How a Scottish Moral Philosopher."

27 Liz Hoffman, "Sam Bankman-Fried, Elon Musk, and a Secret Text," *Semafor*, November 23, 2022, www.semafor.com/article/11/22/2022/sam-bankman-fried-elon-and-a-secret-text. Link there to the texts themselves: "Exhibit H," *Twitter, Inc. v. Elon R. Musk et al.*, C.A. no. 2022-0613-KSJM, September 15, 2022, https://s3.documentcloud.org/documents/23112929/elon-musk-text-exhibits-twitter-v-musk.pdf. EA forum comments: "dyj34650," "Will MacAskill's Role in Connecting SBF to Elon Musk for a Potential Twitter Deal," Effective Altruism Forum, November 12, 2022, https://forum.effectivealtruism.org/posts/dk9HTJKNAAwaEZTgk/will-macaskill-s-role-in-connecting-sbf-to-elon-musk-for-a.

28 "About Us," CSET, accessed June 16, 2024, https://cset.georgetown.edu/about-us/.

29 CSET, "New Grant Agreement Boosts CSET Funding to More Than $100 Million Through 2025," news release, August 24, 2021, https://cset.georgetown.edu/article/new-grant-agreement-boosts-cset-funding-to-more-than-100-million-over-five-years/.

30 "Georgetown University—Center for Security and Emerging Technology," Open Philanthropy, January 2019, www.openphilanthropy.org/grants/georgetown-university-center-for-security-and-emerging-technology; CSET, "New Grant"; "Center for Security and Emerging Technology—General Support (August 2021)," Open Philanthropy, August 2021, www.openphilanthropy.org/grants/center-for-security-and-emerging-technology-general-support-august-2021/; "Center for Security and Emerging Technology—General Support (January 2021)," Open Philanthropy, January 2021, www.openphilanthropy.org/grants/center-for-security-and-emerging-technology-general-support/.

31 Georgetown University, "Q&A with Jason Matheny, Founding Director of the Center for Security and Emerging Technology," news release, February 28, 2019, www.georgetown.edu/news/qa-with-jason-matheny-founding-director-of-the-center-for-security-and-emerging-technology/.

32 Open Philanthropy, "Georgetown University—Center for Security and Emerging Technology."

33 Ibid. Also, "Dr. Jason Matheny," Center for a New American Security, accessed June 16, 2024, www.cnas.org/people/dr-jason-matheny; FHI, "US IARPA Director Jason Matheny Visits," news release, September 30, 2016, www.fhi.ox.ac.uk/us-iarpa-director-jason-matheny-visit/.

34 Jason G. Matheny, "Reducing the Risk of Human Extinction," *Risk Analysis* 27, no. 5 (December 7, 2007): 1335–1344, https://doi.org/10.1111/j.1539-6924.2007.00960.x.

35 Dylan Matthews, "How Effective Altruism Went from a Niche Movement to a Billion-Dollar Force," *Vox*, August 8, 2022, www.vox.com/future-perfect/2022/8/8/23150496/effective-altruism-sam-bankman-fried-dustin-moskovitz-billionaire-philanthropy-crytocurrency.

36 Jason Matheny, "Effective Altruism in Government," Effective Altruism, June 3, 2017, www.effectivealtruism.org/articles/effective-altruism-in-government-jason-matheny. Called an EA Global talk here: Centre for Effective Altruism, "Jason Matheny wanted to do good . . .," Facebook, October 1, 2018, www.facebook.com/CentreforEffectiveAltruism/photos/a.1897756126985429/1923921397702235/?type=3.

37 Doug Irving, "'The Future Could Be Brilliant': RAND's CEO Is an 'Apocaloptimist,'" RAND, August 4, 2022, www.rand.org/pubs/articles/2022/the-future-could-be-brilliant-rands-ceo-is-an-apocaloptimist.html.

38 Bredan Bordelon, "Think Tank Tied to Tech Billionaires Played Key Role in Biden's AI Order," *Politico*, January 16, 2023, www.politico.com/news/2023/12/15/billionaire-backed-think-tank-played-key-role-in-bidens-ai-order-00132128.

39 "RAND Corporation—Emerging Technology Initiatives," Open Philanthropy, October 2023, www.openphilanthropy.org/grants/rand-corporation-emerging-technology-initiatives/; "RAND Corporation—Emerging Technology Fellowships and Research," Open Philanthropy, April 2023, www.openphilanthropy.org/grants/rand-corporation-emerging-technology-fellowships-and-research/; "RAND Corporation—Biosecurity Policy," Open Philanthropy, May 2023, www.openphilanthropy.org/grants/rand-corporation-biosecurity-policy/.

40 Brendan Bordelon, "How a Billionaire-Backed Network of AI Advisers Took Over Washington," *Politico*, October 13, 2023, www.politico.com/news/2023/10/13/open-philanthropy-funding-ai-policy-00121362.

41 "Carrick Flynn," Future of Humanity Institute, archived February 3, 2024, at the Wayback Machine, https://web.archive.org/web/20240203043647/https://www.fhi.ox.ac.uk/team/carrick-flynn/; "Carrick Flynn," CSET, accessed June 16, 2024, https://cset.georgetown.edu/staff/carrick-flynn/; Carrick Flynn, LinkedIn profile, accessed June 16, 2024, www.linkedin.com/in/carrickflynn; "Team," Centre for the Governance of AI, accessed June 16, 2024, www.governance.ai/people; "Carrick Flynn," Centre for the Governance of AI, accessed June 16, 2024, www.governance.ai/team/carrick-flynn; "Carrick Flynn," Brookings, accessed June 16, 2024, www.brookings.edu/people/carrick-flynn/; "Oregon's 6th Congressional District Election, 2022," Ballotpedia, accessed June 16, 2024, https://ballotpedia.org/Oregon%27s_6th_Congressional_District_election,_2022.

42 Dirk VanderHart, "What Does a Crypto Tycoon Want with Oregon's New Congressional District?," Oregon Public Broadcasting, April 24, 2022, www.opb.org/article/2022/04/25/what-does-a-crypto-tycoon-want-with-oregons-new-congressional-district/; Miranda Dixon-Luinenburg and Dylan Matthews, "Carrick Flynn May Be 2022's Unlikeliest Candidate. Here's Why He's Running," *Vox*, May 14, 2022, www.vox.com/23066877/carrick-flynn-effective-altruism-sam-bankman-fried-congress-house-election-2022.

43 Cullen O'Keefe et al., *The Windfall Clause: Distributing the Benefits of AI* (Oxford: FHI, 2020), www.fhi.ox.ac.uk/windfallclause/.

44 Sabrina Rodriguez, "Crypto Gets Its First Big Political Test," *Politico*, May 17, 2022, www.politico.com/news/2022/05/17/oregon-elections-crypto-super-pac-00032960; "Protect Our Future PAC Independent Expenditures," Open Secrets, accessed June 16,

2024, www.opensecrets.org/political-action-committees-pacs/protect-our-future-pac/C00801514/independent-expenditures/2022; "About," Andrea Salinas, House of Representatives, accessed June 16, 2024, https://salinas.house.gov/about.

45 Chris Lehman, "Andrea Salinas, Carrick Flynn Emerge as Frontrunners in Democrats' 6th District Race for Congress," *The Oregonian*, May 5, 2022, www.oregonlive.com/politics/2022/05/andrea-salinas-carrick-flynn-emerge-as-frontrunners-in-democrats-6th-district-race-for-congress.html; Ballotpedia, "Oregon's 6th Congressional."

46 Gabriel Pogrund, "What the Vote Leave Chief Honestly Thinks About Brexit," *New Statesman*, October 20, 2016, www.newstatesman.com/politics/the-staggers/2016/10/what-vote-leave-chief-honestly-thinks-about-brexit; Euan Ritchie and Ian Mitchell, "The UK as an Effective Altruist," Center for Global Development, August 10, 2020, www.cgdev.org/publication/uk-effective-altruist; Press Association, "Millionaire Donors and Business Leaders Back Vote Leave Campaign to Exit EU," *The Guardian*, October 9, 2015, www.theguardian.com/politics/2015/oct/09/millionaire-donors-back-cross-party-campaign-to-leave-eu.

47 "Dominic Cummings: PM's Top Adviser Leaves No 10 to 'Clear the Air,'" BBC, November 13, 2020, www.bbc.com/news/uk-politics-54938050; Alex Wickham, "How Dominic Cummings Took Control in Boris Johnson's First Days as Prime Minister," *BuzzFeed News*, July 27, 2019, www.buzzfeed.com/alexwickham/how-dominic-cummings-took-control-in-boris-johnsons-first; Sean Morrison, "Key Players in Boris Johnson's Election Campaign: From Dominic Cummings to Isaac Levido and Lee Cain," *Standard*, December 13, 2019, www.standard.co.uk/news/politics/boris-johnson-election-campaign-key-players-a4312441.html; Henry Mance and George Parker, "Combative Brexiter Who Took Control of Vote Leave Operation," *Financial Times*, June 13, 2016, www.ft.com/content/cceb7038-30cc-11e6-bda0-04585c31b153#axzz4H22r0hWH.

48 Dominic Cummings, "'Two Hands Are a Lot'—We're Hiring Data Scientists, Project Managers, Policy Experts, Assorted Weirdos . . .," *Dominic Cumming's Blog*, January 2, 2020, https://dominiccummings.com/2020/01/02/two-hands-are-a-lot-were-hiring-data-scientists-project-managers-policy-experts-assorted-weirdos/; Tom Chivers, "Could the Cummings Nerd Army Fix Broken Britain?," *UnHerd*, January 7, 2020, https://unherd.com/2020/01/could-the-cummings-nerd-army-fix-broken-britain/.

49 Peter Walker, Dan Sabbagh, and Rajeev Syal, "Boris Johnson Boots Out Top Adviser Dominic Cummings," *The Guardian*, November 13, 2020, www.theguardian.com/politics/2020/nov/13/dominic-cummings-has-already-left-job-at-no-10-reports; Aubrey Allegretti, "Dominic Cummings Leaves Role with Immediate Effect at PM's Request," SkyNews, November 14, 2020, https://news.sky.com/story/dominic-cummings-leaves-role-with-immediate-effect-at-pms-request-12131792; Mark Landler and Stephen Castle, "'Lions Led by Donkeys': Cummings Unloads on Johnson Government," *New York Times*, May 26, 2021, updated July 18, 2021, www.nytimes.com/2021/05/26/world/europe/cummings-johnson-covid.html.

50 Boris Johnson, "Final Speech as Prime Minister," September 6, 2022, Downing Street, London,www.gov.uk/government/speeches/boris-johnsons-final-speech-as-prime-minister-6-september-2022.

51 *The Guardian*, front page, archived May 26, 2023, at the Wayback Machine, https://web.archive.org/web/20230526150343/theguardian.com/uk.

52 Kiran Stacey and Rowena Mason, "Rishi Sunak Races to Tighten Rules for AI Amid Fears of Existential Risk," *The Guardian*, May 26, 2023, www.theguardian.com/technology/2023/may/26/rishi-sunak-races-to-tighten-rules-for-ai-amid-fears-of-existential-risk.

53 Laurie Clarke, "How Silicon Valley Doomers Are Shaping Rishi Sunak's AI plans," *Politico*, September 14, 2023, www.politico.eu/article/rishi-sunak-artificial-intelligence-pivot-safety-summit-united-kingdom-silicon-valley-effective-altruism/.

54 Toby Ord (@tobyordoxford), Twitter (now X), May 30, 2023, https://twitter.com/tobyordoxford/status/1663550874105581573.

55 Geoffrey Hinton et al., "Statement on AI Risk," Center for AI Safety, May 30, 2023, www.safe.ai/statement-on-ai-risk.

56 "Center for AI Safety—General Support (2023)," Open Philanthropy, April 2023, www.openphilanthropy.org/grants/center-for-ai-safety-general-support-2023/; "Center for AI Safety—Philosophy Fellowship and NeurIPS Prizes," Open Philanthropy, February 2023, www.openphilanthropy.org/grants/center-for-ai-safety-philosophy-fellowship/; "Center for AI Safety—General Support (2022)," Open Philanthropy, November 2022, www.openphilanthropy.org/grants/center-for-ai-safety-general-support/; Jonathan Randles and Steven Church, "FTX Is Probing $6.5 Million Paid to Leading Nonprofit Group on AI Safety," Bloomberg, October 25, 2023, www.bloomberg.com/news/articles/2023-10-25/ftx-probing-6-5-million-paid-to-leading-ai-safety-nonprofit. See also docket no. 3369 from "Ftx Trading Ltd., Case no. 22-11068," November 14, 2022, Kroll, https://restructuring.ra.kroll.com/FTX/Home-DocketInfo.

57 Kevin Roose, "A.I. Poses 'Risk of Extinction,' Industry Leaders Warn," *New York Times*, May 30, 2023, www.nytimes.com/2023/05/30/technology/ai-threat-warning.html; Sheila Chiang, "A.I. Poses Human Extinction Risk on Par with Nuclear War, Sam Altman and Other Tech Leaders Warn," CNBC, May 30, 2023, updated May 31, 2023, www.cnbc.com/2023/05/31/ai-poses-human-extinction-risk-sam-altman-and-other-tech-leaders-warn.html; Matt O'Brien, "Artificial Intelligence Raises Risk of Extinction, Experts Say in New Warning," AP, May 30, 2023, https://apnews.com/article/artificial-intelligence-risk-of-extinction-ai-54ea8aadc60d1503e5a65878219aad43; Max Zahn, "AI Leaders Warn the Technology Poses 'Risk of Extinction' Like Pandemics and Nuclear War," ABC News, May 30, 2023, https://abcnews.go.com/Technology/ai-leaders-warn-technology-poses-risk-extinction-pandemics/story?id=99690874; James Vincent, "Top AI Researchers and CEOs Warn Against 'Risk of Extinction' in 22-Word Statement," *The Verge*, May 30, 2023, www.theverge.com/2023/5/30/23742005/ai-risk-warning-22-word-statement-google-deepmind-openai; Vanessa Romo, "Leading Experts Warn of a Risk of Extinction from AI," NPR, May 30, 2023, www.npr.org/2023/05/30/1178943163/ai-risk-extinction-chatgpt; Chris Vallance, "Artificial Intelligence Could Lead to Extinction, Experts Warn," BBC, May 30, 2023, www.bbc.com/news/uk-65746524; Aaron Gregg, Cristiano Lima-Strong, and Gerrit De Vynck, "AI Poses 'Risk of Extinction' on Par with Nukes, Tech Leaders Say," *Washington Post*, May 30, 2023, www.washingtonpost.com/business/2023/05/30/ai-poses-risk-extinction-industry-leaders-warn/.

58 Hinton et al., "Statement on AI Risk"; "Who We Are," Future Fund, accessed June 16, 2024, https://ftxfuturefund.org.cach3.com/index.html%3Fp=32.html; "Future Fund," Effective Altruism Forum, last modified June 21, 2023, https://forum.effectivealtruism.org/topics/future-fund; Nicholas Kulish, "FTX's Collapse Casts a Pall on a Philanthropy Movement," *New York Times*, November 13, 2022, updated November 14, 2022, www.nytimes.com/2022/11/13/business/ftx-effective-altruism.html.

59 Ord interview.

60 Ibid. Here's one of the surveys he's talking about: Zach Stein-Perlman and Katja Grace, "2022 Expert Survey on Progress in AI," AI Impacts, last modified May 26, 2023, https://wiki.aiimpacts.org/doku.php?id=ai_timelines:predictions_of_human-level_ai_timelines:ai_timeline_surveys:2022_expert_survey_on_progress_in_ai.

61 Melanie Mitchell, "Do Half of AI Researchers Believe That There's a 10% Chance AI Will Kill Us All?," *AI: A Guide for Thinking Humans* (blog), April 23, 2023, https://aiguide.substack.com/p/do-half-of-ai-researchers-believe.

62 Mitchell interview.

63 Ibid.

64 Ord interview.

65 Nick Bostrom, "Existential Risk Prevention as Global Priority," *Global Policy* 4, no. 1 (2013): 15–31, https://existential-risk.com/concept.

66 William MacAskill and Hilary Greaves, "The Case for Strong Longtermism" (working paper, Global Priorities Institute, June 2021), https://globalprioritiesinstitute.org/wp-content/uploads/The-Case-for-Strong-Longtermism-GPI-Working-Paper-June-2021-2-2.pdf. Their estimate of lives saved with malaria bed net distribution comes from GiveWell, and is about 0.025 lives per $100 spent (ibid., 5).

67 "About Me," NickBeckstead.com, accessed June 16, 2024, www.nickbeckstead.com/.

68 Nicholas Beckstead, "On the Overwhelming Importance of Shaping the Far Future" (PhD diss., Rutgers University, May 2013), https://doi.org/doi:10.7282/T35M649T.

69 Ord, *Precipice*, 46, 286, 306.

70 Beckstead, "On the Overwhelming Importance," 11.

71 This is found on page 11 of the version of his thesis linked to on Beckstead's website: Beckstead, "On the Overwhelming Importance," via "Research," NickBeckstead.com, accessed June 16, 2024, https://drive.google.com/file/d/0B8P94pg6WYCIc0lXSUVYS1BnMkE/view?resourcekey=0-nk6wM1QIPl0qWVh2z9FG4Q.

72 Ibid.

73 Dylan Matthews, "I Spent a Weekend at Google Talking with Nerds About Charity. I Came Away . . . Worried," *Vox*, August 10, 2015, www.vox.com/2015/8/10/9124145/effective-altruism-global-ai.

74 Thorstad, interview with author, May 29, 2023.

75 Ord interview.

76 Ibid.

77 Ord, *Precipice*, 11.

78 Ibid., p. 52.

79 Thorstad interview, May 29.

80 Thorstad, interview with author, April 26, 2023.

81 Brian Weatherson (@bweatherson), Twitter, August 9, 2022, archived August 9, 2022, at the Wayback Machine, https://web.archive.org/web/20220809194042/https://twitter.com/bweatherson/status/1557035220855099392.

82 Ord interview.

83 Thorstad interview, May 29.

84 Thorstad interview, April 26.

85 David Thorstad, "Existential Risk Pessimism and the Time of Perils (Part 9: Objections and Replies)," *Reflective Altruism* (blog), April 1, 2023, https://ineffectivealtruismblog.com/2023/04/01/existential-risk-pessimism-and-the-time-of-perils-part-9-objections-and-replies/.

86 Thorstad interview, April 26. The post is here: Carl Shulman, comment on "The Discount Rate Is Not Zero," Effective Altruism Forum, September 3, 2022, https://forum.effectivealtruism.org/posts/zLZMsthcqfmv5J6Ev/?commentId=Nr35E6sTfn9cPxrwQ. And here's an example of that comment being used as a standard reference among EAs: "Time of Perils," Effective Altruism Forum, last modified October 1, 2022, https://forum.effectivealtruism.org/topics/time-of-perils.

87 Thorstad interview, May 29.

88 Ord, *Precipice*, 227–231.

89 Ibid., 231.

90 MacAskill, *What We Owe*, 189.

91 Timothy Ferris, *The Creation of the Universe* (PBS/Northstar Productions, 1985). See also, Nicholos Wethington, "The Switch to Digital Switches Off Big Bang TV Signal," *Universe Today*, February 16, 2009, www.universetoday.com/25560/the-switch-to-digital-switches-off-big-bang-tv-signal/; Hugh Lippincott, "Understanding the CMB," *Lippincott Lab*, August 1, 2010, http://hep.ucsb.edu/people/hugh/2010/08/01/test2/.

92 Fred C. Adams and Gregory Laughlin, "A Dying Universe: The Long-Term Fate and Evolution of Astrophysical Objects," *Reviews of Modern Physics* 69, no. 337 (April 1, 1997), https://journals.aps.org/rmp/abstract/10.1103/RevModPhys.69.337.

93 Nick Bostrom, "Astronomical Waste: The Opportunity Cost of Delayed Technological Development," *Utilitas* 15, no. 3 (2003): 308–314, https://nickbostrom.com/astronomical/waste (emphasis in the original).

94 Ord, *Precipice*, 233–235.

95 Stuart Armstrong and Anders Sandberg, "Eternity in Six Hours: Intergalactic Spreading of Intelligent Life and Sharpening the Fermi Paradox," *Acta Astronautica* 89 (August–September 2013): 1–13, https://doi.org/10.1016/j.actaastro.2013.04.002.

96 Ibid.

97 Anders Sandberg, Eric Drexler, and Toby Ord, "Dissolving the Fermi Paradox," arXiv:1806.02404v1 (June 6, 2018), https://doi.org/10.48550/arXiv.1806.02404.

98 Annalee Newitz, *Scatter, Adapt, and Remember: How Humans Will Survive a Mass Extinction* (New York: Doubleday, 2013), 252.

99 There's also a question of how to trust the AGIs in the probes themselves. Even if we grant all the different steps in the argument about the importance of AI alignment, and then further grant the future existence of an aligned AGI that builds these probes and populates them with its electronic progeny, there does seem to be a serious risk in creating many copies of AGIs. No copying process can be perfect, especially in a high-radiation environment in space. Eventually, a bad copy might be made that falls out of alignment, with a massive arsenal ready at hand. I don't think this is a good argument, because I don't buy any of the unlikely premises at work in this entire scheme, but this does seem like the kind of thing that the longtermists and rationalists should be more concerned about, by their own logic.

100 Armstrong and Sandberg, "Eternity in Six Hours."

101 Carl Sagan, *Cosmos* (New York: Random House, 1980), 130.

102 William MacAskill, "Afterwards," in *What We Owe*, posted to WhatWeOwetheFuture.com, accessed June 16, 2024, https://whatweowethefuture.com/afterwards/.

103 Kieran Setiya, "The New Moral Mathematics," *Boston Review*, August 15, 2022, www.bostonreview.net/articles/the-new-moral-mathematics/.

104 Ibid.

105 MacAskill, *What We Owe*, 167–168.

106 Ibid., 180–181.

107 Stéphane Zuber et al., "What Should We Agree on About the Repugnant Conclusion?," *Utilitas* 33, no. 4 (December 2021):379–383, https://doi.org/10.1017/S095382082100011X.

108 MacAskill, *What We Owe*, 184.

109 See Ursula K. Le Guin, "The Ones Who Walk Away from Omelas," in *The Wind's Twelve Quarters* (New York: Harper & Row, 1975).

110 Toby Ord, "Why I'm Not a Negative Utilitarian," *A Mirror Clear* (blog), February 28, 2013, updated March 1, 2013, www.amirrorclear.net/academic/ideas/negative-utilitarianism/ (emphasis in the original).

111 Torres, interview with author, July 26, 2023.

112 Setiya, "New Moral Mathematics."

113 Ibid.

114 MacAskill, *What We Owe*, 186.

115 Tyler Cowen, "William MacAskill on Effective Altruism, Moral Progress, and Cultural Innovation (Ep. 156)," August 10, 2022, in *Conversations with Tyler*, podcast, https://conversationswithtyler.com/episodes/william-macaskill/.

116 MacAskill, *What We Owe*, 186–187.

117 Setiya, "New Moral Mathematics."

118 MacAskill, "Afterwards" (emphasis in the original).

119 Setiya, "New Moral Mathematics."

120 MacAskill, "Afterwards."

121 Torres, interview with author, July 18, 2023.

122 Isaiah Berlin, "The Pursuit of the Ideal," in *The Crooked Timber of Humanity* (Princeton, NJ: Princeton University Press, 2013).

123 Amia Srinivasan, "Stop the Robot Apocalypse," *London Review of Books* 37, no. 18 (September 24, 2015), www.lrb.co.uk/the-paper/v37/n18/amia-srinivasan/stop-the-robot-apocalypse.

124 Ibid.

125 Reich interview.

126 Ibid.

127 Srinivasan, "Stop the Robot Apocalypse."

128 Gerard, interview with author.

CHAPTER 5

1 Marc Andreessen, "The Techno-Optimist Manifesto," Andreessen Horowitz, October 16, 2023, https://a16z.com/the-techno-optimist-manifesto/.

2 "A Short History of the Web," CERN, accessed June 13, 2024, https://home.cern/science/computing/birth-web/short-history-web.

3 Jay Hoffman, "The Origin of the IMG Tag," *History of the Web*, March 7, 2017, https://thehistoryoftheweb.com/the-origin-of-the-img-tag/.

4 Robert Metcalfe, "Microsoft and Netscape Open Some New Fronts in Escalating Web Wars," *InfoWorld*, August 21, 1995, p. 35, https://archive.org/details/bub_gb_0joEAAAAMBAJ/page/n42/mode/1up.

5 Andreessen, "Manifesto."

6 Ibid. (emphasis his).

7 Emily Baker-White, "Who Is @BasedBeffJezos, the Leader of the Tech Elite's 'E/acc' Movement?," *Forbes*, December 1, 2023, www.forbes.com/sites/emilybaker

-white/2023/12/01/who-is-basedbeffjezos-the-leader-of-effective-accelerationism -eacc/.

8 "swarthy," "Effective Accelerationism—e/acc," *swarthy's sensibles* (blog), May 31, 2022, archived June 2, 2022, at the Wayback Machine, https://web.archive.org /web/20220602164803/https://swarthy.substack.com/p/effective-accelerationism -eacc?s=r. Republished as "zestular" et al., "Effective Accelerationism — E/acc," *E/acc Newsletter*, October 31, 2022, https://effectiveaccelerationism.substack.com/p /repost-effective-accelerationism.

9 "What the F* Is E/acc," *E/acc Newsletter*, December 26, 2022, https://effectiveaccel erationism.substack.com/p/what-the-f-is-eacc. See also, "Notes on E/acc Principles and Tenets," *E/acc Newsletter*, October 31, 2022, https://effectiveaccelerationism.substack .com/p/repost-notes-on-eacc-principles-and.

10 "swarthy," "Effective Accelerationism."

11 The effective accelerationists are correct about one thing: life is a dissipative process. Life intercepts entropy flows, turning low-entropy sources of energy into higher entropy. But life isn't the only dissipative process; lots of things do that. In fact, every process that actually happens in the universe does exactly that. Life is far from the most efficient of those processes. The most efficient way to generate entropy from a source of energy is to drop it into a black hole. So, insofar as the universe has a "thermodynamic will," it's to drop things into black holes. The effective accelerationists should be agitating for the construction of a fleet of black holes to throw things into.

This is part of a bigger problem for them, a very old philosophical chestnut: the difference between "is" and "ought." Entropy is increasing, but that doesn't mean that we ought to increase entropy as much as we can. Indeed, there are good and simple arguments that we should not do that. Consider a variant on the classic trolley problem: A trolley is heading toward a fork in the tracks. If you do nothing, the trolley will take the right-hand track, killing five people who are on the tracks and can't get away in time. If instead you pull the control lever, the trolley will take the left-hand track, which leads to the black-hole factory. There, the trolley will destroy the safety mechanisms in place and thereby create a new black hole that swallows the entire Earth. If the effective accelerationists truly believe that they must be agents of the "thermodynamic will of the universe," the choice is easy: pulling the lever creates far more entropy than not pulling it. While the effective accelerationists have certainly taken a novel position on what constitutes moral actions, this does not appear to be a tenable theory of ethics.

12 "Notes on E/acc Principles and Tenets."

13 Altman, "Moore's Law for Everything."

14 Tyler Cowen, "Sam Altman on Loving Community, Hating Coworking, and the Hunt for Talent (Ep. 61—Live in SF)," February 27, 2019, in *Conversations with Tyler*, podcast, https://conversationswithtyler.com/episodes/sam-altman/.

15 Tad Friend, "Sam Altman's Manifest Destiny," *New Yorker*, October 3, 2016, www .newyorker.com/magazine/2016/10/10/sam-altmans-manifest-destiny.

16 "Chat with OpenAI CEO and Co-founder Sam Altman, and Chief Scientist Ilya Sutskever," video, Tel Aviv University, YouTube, June 5, 2023, www.youtube.com

/watch?v=mC-0XqTAeMQ. Source on date of talk: Tel Aviv University, "Sam Altman to Visit Tel Aviv University," news release, May 31, 2023, https://english.tau.ac.il/news/sam_altman_tau.

17 Brad Stone and Sam Altman, "Open AI's Sam Altman: America Will Be Fine After the Election," video, Bloomberg House Davos, January 16, 2024, 23:22, www.bloomberg.com/news/videos/2024-01-16/openai-s-atlman-and-makanju-on-global-implications-of-ai. See also Paris Marx, "Sam Altman's Self-Serving Vision of the Future," *Disconnect* (blog), January 24, 2024, https://disconnect.blog/sam-altmans-self-serving-vision-of-the-future/.

18 John Vidal, "Bill Gates Backs Climate Scientists Lobbying for Large-Scale Geoengineering," *The Guardian*, February 6, 2012, www.theguardian.com/environment/2012/feb/06/bill-gates-climate-scientists-geoengineering; Corbin Hiar and E&E News, "Solar Geoengineering Looks to Silicon Valley for New Wave of Funding," *Scientific American*, February 15, 2024, www.scientificamerican.com/article/solar-geoengineering-looks-to-silicon-valley-for-new-wave-of-funding/; Chelsea Harvey and E&E News, "Geoengineering Is Not a Quick Fix for the Climate Crisis, New Analysis Shows," *Scientific American*, April 3, 2023, www.scientificamerican.com/article/geoengineering-is-not-a-quick-fix-for-the-climate-crisis-new-analysis-shows/; Paris Marx, "We Don't Need New Tech to Fight Climate Change," *Disconnect* (blog), March 23, 2023, https://disconnect.blog/we-dont-need-more-tech-to-fight-climate/.

19 Mark Zuckerberg, Tyler Cowen, and Patrick Collison, "Mark Zuckerberg Interviews Patrick Collison and Tyler Cowen on the Nature and Causes of Progress (BONUS)," November 22, 2019, in *Conversations with Tyler*, podcast, https://conversationswithtyler.com/episodes/mark-zuckerberg-interviews-patrick-collison-and-tyler-cowen/, transcript taken from https://patrickcollison.com/static/files/progress-mz.pdf.

20 Patrick Collison and Tyler Cowen, "We Need a New Science of Progress," *The Atlantic*, July 30, 2019, www.theatlantic.com/science/archive/2019/07/we-need-new-science-progress/594946/.

21 Jason Crawford, "The Roots of Progress Is Now a Nonprofit Organization," *Roots of Progress* (blog), Roots of Progress Institute, August 23, 2021, https://rootsofprogress.org/nonprofit-announcement; "About," Roots of Progress Institute, accessed June 13, 2024, https://rootsofprogress.org/about.

22 "This Is Works in Progress," *Works in Progress*, accessed June 13, 2024, https://worksinprogress.co/about/.

23 Benjamin Reinhardt, "Making Energy Too Cheap to Meter," *Works in Progress*, October 12, 2022, https://worksinprogress.co/issue/making-energy-too-cheap-to-meter/(emphasis his).

24 Ben Reinhardt, "About," BenjaminReinhardt.com, accessed June 13, 2024, https://benjaminreinhardt.com/about/.

25 Reinhardt, "Making Energy."

26 Reinhardt, interview with author.

27 Reinhardt, "Making Energy."

28 Reinhardt interview.

29 Trang Chu Minh, "Five Facts About Unsustainable Waste Management in Singapore," *Eco-Business*, January 11, 2021, www.eco-business.com/opinion/five-facts-about-unsustainable-waste-management-in-singapore/; Robin Hicks, "How Will Singapore Defuse a 16-Year Waste Timebomb?," *Eco-Business*, May 25, 2019, archived May 25, 2019, at the Wayback Machine, https://web.archive.org/web/20190525082946/https://www.eco-business.com/news/how-will-singapore-defuse-a-16-year-waste-timebomb/; Martin Abbugao, "Singapore Races to Save 'Garbage of Eden,'" Phys.org, July 28, 2023, https://phys.org/news/2023-07-singapore-garbage-eden.html.

30 Reinhardt, "Making Energy."

31 Ibid.

32 Hall, *Flying Car*, 169.

33 Ibid., 28.

34 Ibid., 271, 34.

35 Ibid., 288.

36 Ibid., 101.

37 Ibid., 104 (emphasis his).

38 Lex Fridman, "Jeff Bezos: Amazon and Blue Origin | Lex Fridman Podcast #405," December 14, 2023, in *Lex Fridman Podcast*, transcript, https://lexfridman.com/jeff-bezos-transcript.

39 Mosher, "Jeff Bezos Just Gave a Private Talk in New York."

40 Hall, *Flying Car*, 249.

41 Reinhardt interview.

42 Jordan Robertson, "Elon Musk's Vision: 'I Would Like to Die on Mars. Just Not on Impact,'" Yahoo Finance, September 26, 2023, https://finance.yahoo.com/news/elon-musks-vision-die-mars-165604970.html.

43 Richard Heinberg, "The Limits to Growth at 50: From Scenarios to Unfolding Reality," *Resilience*, February 24, 2022, www.resilience.org/stories/2022-02-24/the-limits-to-growth-at-50-from-scenarios-to-unfolding-reality/.

44 Donella H. Meadows et al., *The Limits to Growth: A Report for the Club of Rome's Project on the Predicament of Mankind* (New York: Universe Books, 1972), 29.

45 McCray, *Visioneers*, 37.

46 Ibid., 36–37.

47 Ibid., 38.

48 Meadows et al., *Limits to Growth*, 159.

49 Eliezer Yudkowsky, "Raised in Technophilia," LessWrong, September 16, 2008, www.lesswrong.com/posts/uNWRXtdwL33ELgWjD/raised-in-technophilia.

50 Daniel Deudney, *Dark Skies: Space Expansionism, Planetary Geopolitics, and the Ends of Humanity* (Oxford: Oxford University Press, 2020), 200–202.

51 McCray, *Visioneers*, 48.

52 Richard Fisher, "The Long-Term Quest to Build a 'Galactic Civilisation,'" BBC, July 21, 2021, www.bbc.com/future/article/20210721-the-quest-for-a-galactic-civilisation-that-saves-humanity.

53 Franklin Foer, "Jeff Bezos's Master Plan," *The Atlantic*, November 2019, www.theatlantic.com/magazine/archive/2019/11/what-jeff-bezos-wants/598363/.

54 Caroline Haskins, "Jeff Bezos Is a Post-Earth Capitalist," *Motherboard*, May 10, 2019, www.vice.com/en/article/3k3kwb/jeff-bezos-is-a-post-earth-capitalist.

55 Fridman, "Bezos."

56 Döpfner, "Jeff Bezos Interview."

57 Fridman, "Bezos."

58 Foer, "Jeff Bezos's Master Plan."

59 McCray, *Visioneers*, 56–59; Mike Wall, "Artemis 1 Launch Success Makes NASA's SLS the Most Powerful Rocket Ever to Fly," Space.com, November 16, 2022, www.space.com/nasa-artemis-1-sls-most-powerful-rocket; NASA, "Space Launch System," accessed June 13, 2024, www.nasa.gov/wp-content/uploads/2021/04/0080_sls_fact_sheet_sept2020_09082020_final_0.pdf?emrc=d5d3e0.

60 Elon Musk (@elonmusk), Twitter (now X), June 5, 2022, https://x.com/elonmusk/status/1533410745429413888; Victor Tangermann, "Elon Musk: We Have to Get to Mars Because of 'Nuclear Armageddon' on Earth," *Futurism*, November 11, 2021, https://futurism.com/elon-musk-nuclear-armageddon.

61 Elon Musk (@elonmusk), Twitter (now X), June 24, 2018, https://twitter.com/elonmusk/status/1011083630301536256. See also, Nick Lucchesi, "Elon Musk Calls on the Public to 'Preserve Human Consciousness' with Starship," *Inverse*, September 28, 2019, www.inverse.com/article/59676-spacex-starship-presentation.

62 George Dvorsky, "Elon Musk's Plan to Send a Million Colonists to Mars by 2050 Is Pure Delusion," *Gizmodo*, June 3, 2022, https://gizmodo.com/elon-musk-mars-colony-delusion-1848839584.

63 Elon Musk (@elonmusk), Twitter (now X), March 15, 2024, https://twitter.com/elonmusk/status/1768810190718009446.

64 Kenneth Chang, "Elon Musk Sets Out SpaceX Starship's Ambitious Launch Timeline," *New York Times*, September 28, 2019, updated May 5, 2021, www.nytimes.com/2019/09/28/science/elon-musk-spacex-starship.html.

65 Mrigakshi Dixit, "How Space Radiation Threatens Lunar Exploration," *Smithsonian Magazine*, January 18, 2023, www.smithsonianmag.com/science-nature/how-space-radiation-threatens-lunar-exploration-180981415/.

66 "Steady Temperatures at Mars' Gale Crater," NASA, April 8, 2013, https://science.nasa.gov/resource/steady-temperatures-at-mars-gale-crater/.

67 Carl Zimmer, "The Lost History of One of the World's Strangest Science Experiments," *New York Times*, March 29, 2019, www.nytimes.com/2019/03/29/sunday-review/biosphere-2-climate-change.html.

68 Taylor Locke, "Elon Musk on Planning for Mars: 'The City Has to Survive if the Resupply Ships Stop Coming from Earth,'" CNBC, March 9, 2020, updated January 12, 2021, www.cnbc.com/2020/03/09/spacex-plans-how-elon-musk-see-life-on-mars.html; Eric Berger, "Inside Elon Musk's Plan to Build One Starship a Week—and Settle Mars," *Ars Technica*, March 5, 2020, https://arstechnica.com/science/2020/03/inside-elon-musks-plan-to-build-one-starship-a-week-and-settle-mars/; Mike Brown, "SpaceX Mars City: Elon Musk Details 1 Test Its Success Depends On," *Inverse*, October 19, 2020, updated February 20, 2024, www.inverse.com/innovation/spacex-mars-city-test.

69 Paul Krugman, "Elon Musk, Mars and the Modern Economy," *New York Times*, June 7, 2022, www.nytimes.com/2022/06/07/opinion/musk-mars-twitter.html; Charles Stross, "Insufficient Data," *Charlie's Diary* (blog), July 23, 2010, www.antipope.org/charlie/blog-static/2010/07/insufficient-data.html.

70 Elon Musk, "Elon Musk Might Be a Super Villain," interview by Stephen Colbert, *Late Show with Stephen Colbert*, video, YouTube, September 10, 2015, www.youtube.com/watch?v=gV6hP9wpMW8.

71 Andrew Ross Sorkin, "Elon Musk and Y Combinator President on Thinking for the Future—FULL CONVERSATION," video, *Vanity Fair*, YouTube, October 8, 2015, 9:40, www.youtube.com/watch?v=SqEo107j-uw.

72 Mike Wall, "Elon Musk Floats 'Nuke Mars' Idea Again (He Has T-Shirts)," Space.com, August 17, 2019, www.space.com/elon-musk-nuke-mars-terraforming.html.

73 John Wenz, "Sorry, Elon. There's Not Enough CO2 to Terraform Mars," *Discover*, July 30, 2018, www.discovermagazine.com/the-sciences/sorry-elon-theres-not-enough-co2-to-terraform-mars; John Wenz, "No Seriously, Elon. You Can't Just Nuke Mars (We Asked)," *Discover*, July 31, 2018, www.discovermagazine.com/the-sciences/no-seriously-elon-you-cant-just-nuke-mars-we-asked.

74 George Dvorsky, "Humans Will Never Colonize Mars," *Gizmodo*, July 30, 2019, https://gizmodo.com/humans-will-never-colonize-mars-1836316222; Shannon Stirone, "Mars Is a Hellhole," *The Atlantic*, February 26, 2021, www.theatlantic.com/ideas/archive/2021/02/mars-is-no-earth/618133/.

75 "Weather on the Moon," NASA, accessed June 13, 2024, https://science.nasa.gov/moon/weather-on-the-moon/.

76 Kelly Weinersmith and Zach Weinersmith, *A City on Mars: Can We Settle Space, Should We Settle Space, and Have We Really Thought This Through?* (New York: Penguin, 2023), 123.

77 Adam Becker, "The Amazing Cloud Cities We Could Build on Venus," BBC, October 20, 2016, www.bbc.com/future/article/20161019-the-amazing-cloud-cities-we-could-build-on-venus.

78 We can't go faster than the speed of light, despite myths, rumors, and hopes to the contrary. Here are answers to a few of the most common objections to the light-speed limit:

- No, the light barrier isn't like the sound barrier. Nobody ever thought that was a truly fixed physical limit. Indeed, humans have had technology that could break the speed of sound for millennia: the crack of a whip is actually a tiny sonic boom, produced as the tail end of the whip goes faster than the speed of sound. Going faster than the speed of light would require fundamentally new physics, not just new engineering.
- Yes, the light barrier has been tested in the lab. Particle accelerators like the Large Hadron Collider at CERN regularly send subatomic particles at speeds very close to the speed of light, and their behavior is always in perfect accordance with relativity. They can't go faster.
- No, quantum entanglement does not give us a way to travel or communicate faster than the speed of light. This is beyond the scope of this book, but for a good introduction to the subject, try my last book, *What Is Real?* (New York: Basic Books, 2018).
- No, general relativity does not allow for the construction of a warp drive. There is a "warp" solution to the Einstein field equation, but it doesn't work like the warp drive on *Star Trek*, and that same equation says it can't be built if it doesn't already exist.
- No, we can't build wormholes. Wormholes are far and away the most promising avenue for going faster than light, and they're almost certainly impossible to use for that. They might not even exist at all; nobody has ever seen one.

We could be surprised. We could be wrong. But given the overwhelming evidence at hand, it would be foolish to expect that a means of traveling faster than light will be discovered, especially anytime soon.

79 Most of the mass needed will be water. At a population of eight billion, humanity's annual water usage is about four trillion cubic meters (four trillion metric tons) a year, not including what's used for saltwater fishing. That comes out to a bit more than a metric ton of water per person per day. I'll be generous and assume that these cylinders are incredibly efficient and entirely recycle their water each day, meaning they only need a day's worth of water for each resident. Therefore, the amount of water needed for a trillion humans would be about a trillion metric tons. A typical comet has a mass of five billion metric tons, and not all of that is water. Thus, hundreds of comets would be needed, at an absolute minimum. (A more realistic figure is probably orders of magnitude higher.) Sources: Hannah Ritchie and Max Roser, "Water Use and Stress," Our World in Data, 2018, updated February 2024, https://ourworldindata.org/water-use-stress; NASA Hubble Mission Team, "Hubble Confirms Largest Comet Nucleus Ever Seen," NASA, April 12, 2022, https://science.nasa.gov/missions/hubble/hubble-confirms-largest-comet-nucleus-ever-seen/. As the article indicates, more massive comets are found farther out in the solar system, but they'd lose much of their mass by the time they got closer to the Sun, where the cylinders would have to be stationed. (It would be hard to get to comets that far out anyhow.)

80 Larry Page, "Where's Google Going Next? | Larry Page," interview by Charlie Rose, recorded at TED2014, British Columbia, Canada, video, YouTube, March 22, 2014, 20:30, www.youtube.com/watch?v=mArrNRWQEso.

81 Brannen, *Ends of the World*, 188.

82 *Encyclopaedia Britannica Online*, s.v. "K–T extinction," May 29, 2024, www.britannica.com/science/K-T-extinction.

83 Peter Brannen (@PeterBrannen1), Twitter (now X), December 10, 2018, https://x.com/PeterBrannen1/status/1072174222967898112.

84 Ibid.

85 Dvorsky, "Humans Will Never Colonize Mars."

86 This is the speed that the *Apollo* missions reached on hitting Earth's atmosphere: W. David Woods, Kenneth D. MacTaggart, and Frank O'Brien, "Apollo 11 Day 9, Part 2: Entry and Splashdown," Apollo Flight Journal, last modified March 3, 2021, www.nasa.gov/history/afj/ap11fj/27day9-entry.html. This is also roughly the speed that any object leaving the moon at its escape velocity would reach after falling from the Moon's orbit down to Earth's atmosphere—just as the *Apollo* missions did.

87 Mike Brown, "SpaceX Mars City: Legal Experts Respond to 'Gibberish' Free Planet Claim," *Inverse*, November 3, 2020, updated February 20, 2024, www.inverse.com/innovation/spacex-mars-city-legal.

88 Ibid.

89 For much, much more on the ways developing space technology can make the world less safe, see Deudney's excellent *Dark Skies*. For more on space law, see Weinersmith and Weinersmith, *City on Mars*, part 4, as well as Adam Mann, "Who's in Charge of Outer Space?," *Wall Street Journal*, May 19, 2017, archived June 1, 2017, at the Wayback Machine, https://web.archive.org/web/20170601130235/www.wsj.com/articles/whos-in-charge-of-outer-space-1495195097.

90 Walkowicz interview.

91 "It Is Difficult to Get a Man to Understand Something When His Salary Depends Upon His Not Understanding It," Quote Investigator, November 30, 2017, https://quoteinvestigator.com/2017/11/30/salary/?amp=1.

92 Loren Grush, "Jeff Bezos' Space Company Is Pressuring Employees to Launch a Tourist Rocket During the Pandemic," *The Verge*, April 2, 2020, www.theverge.com/2020/4/2/21198272/blue-origin-coronavirus-leaked-audio-test-launch-workers-jeff-bezos; Ronan Farrow, "Elon Musk's Shadow Rule," *New Yorker*, August 21, 2023, www.newyorker.com/magazine/2023/08/28/elon-musks-shadow-rule.

93 Elon Musk (@elonmusk), Twitter (now X), May 1, 2020, https://twitter.com/elonmusk/status/1256347698120159233.

94 Elon Musk (@elonmusk), Twitter (now X), July 12, 2021, https://twitter.com/elonmusk/status/1414782972474048516.

95 Walkowicz interview.

96 Ibid.

97 Sarah Knapton, "Human Race Is Doomed if We Do Not Colonise the Moon and Mars, Says Stephen Hawking," *Telegraph*, June 20, 2017, www.telegraph.co.uk/science/2017/06/20/human-race-doomed-do-not-colonise-moon-mars-says-stephen-hawking/.

98 Hall, *Flying Car*, 107.

99 Andreessen, "Manifesto" (emphasis his).

100 Filippo Tommaso Marinetti, "The Futurist Manifesto," 1909, Obelisk, accessed June 13, 2024, www.arthistoryproject.com/artists/filippo-tommaso-marinetti/the-futurist-manifesto/.

101 "swarthy," "Effective Accelerationism." The Land quote is at the very beginning of the post: "Accelerationism is simply the self-awareness of capitalism, which has scarcely begun. ('We haven't seen anything yet.')"

102 Aarti Shahani, "Colonialism Comment Puts Facebook Under Scrutiny," *All Tech Considered*, NPR, February 12, 2016, www.npr.org/sections/alltechconsidered/2016/02/12/466506966/colonialism-comment-puts-facebook-under-scrutiny.

103 Andreessen, "Manifesto."

104 This is a paraphrase of Asimov: "Isaac Asimov Laments the 'Cult of Ignorance' in the United States (1980)," Open Culture, October 12, 2016, www.openculture.com/2016/10/isaac-asimov-laments-the-cult-of-ignorance-in-the-united-states.html; Isaac Asimov, "A Cult of Ignorance," *Newsweek*, January 21, 1980, p. 19.

105 Andreessen, "Manifesto." See also "dril" (@dril), Twitter (now X), December 28, 2014, https://x.com/dril/status/549425182767861760.

106 Jon Bois, "'I WISH EVERYONE ELSE WAS DEAD.' PRETTY GOOD, EPISODE 7," in *Pretty Good*, video, YouTube, April 27, 2016, 19:27, www.youtube.com/watch?v=_P52G4Kyq5M.

107 Hall, interview with author.

CHAPTER 6

1 Kate Crawford, *Atlas of AI: Power, Politics, and the Planetary Costs of Artificial Intelligence* (New Haven, CT: Yale University Press, 2021), 234.

2 McCray, *Visioneers*, 100, 110.

3 Ibid., 196.

4 Regis, *Great Mambo Chicken*, 84–87.

5 Meghan O'Gieblyn, *God, Human, Animal, Machine: Technology, Metaphor, and the Search for Meaning* (New York; Doubleday, 2021), 8.

6 Ibid., 44.

7 Ibid., 57.

8 Ibid., 67.

9 Ibid., 68.

10 George M. Young, *The Russian Cosmists: The Esoteric Futurism of Nikolai Fedorov and His Followers* (Oxford: Oxford University Press, 2012), 81–82 (emphasis in original).

11 Ibid., 60, 73, 78–79, 89.

12 Ibid., 47–51.

13 Fred Scharmen, *Space Forces: A Critical History of Life in Outer Space* (New York: Verso, 2021), 18; Boris Groys, ed., *Russian Cosmism* (Cambridge, MA: MIT Press), 229–230.

14 Young, *Russian Cosmists*, 146.

15 Ibid., 147.

16 Ibid., 147–148.

17 Ibid., 74; Konstantin Tsiolkovsky, "Exploration of the Universe with Reaction Machines," in *Exploring the Unknown: Selected Documents in the History of the U.S. Civil Space Program*, vol. 1, *Organizing for Exploration*, ed. John Lodgson (Washington, DC: NASA History Office, 1995), 59–132, https://web.stanford.edu/~cantwell/AA283_Course_Material/AA283_Resources/Tsiolkovsky,%20Exploration%20of%20the%20Universe%20with%20Reaction%20Machines%201898.pdf.

18 "Konstantin Tsiolkovsky," European Space Agency, accessed June 12, 2024, www.esa.int/Science_Exploration/Human_and_Robotic_Exploration/Exploration/Konstantin_Tsiolkovsky.

19 Scharmen, *Space Forces*, 37.

20 Young, *Russian Cosmists*, 156. See also Adam Frank, *Light of the Stars: Alien Worlds and the Fate of the Earth* (New York: W. W. Norton, 2018), 221, 250; David Ronfeldt and John Arquilla, "Origins and Attributes of the Noosphere Concept," in *Whose Story Wins: Rise of the Noosphere, Noopolitik, and Information-Age Statecraft* (Santa Monica, CA: RAND Corporation, 2020), 7–12, www.jstor.org/stable/resrep26549.9.

21 O'Gieblyn, *God, Human, Animal, Machine*, 71.

22 Ibid.

23 Ibid., 53, says this is the first appearance of "transhuman" in English. For the claim that this is one of the first translations of the *Divine Comedy* into English—specifically, the second complete translation, only a few years after the first—see this surprisingly specific Wikipedia page: Wikipedia, s.v. "List of English translations of the *Divine Comedy*," last modified July 18, 2024, at 21:58, https://en.wikipedia.org/wiki/List_of_English_translations_of_the_Divine_Comedy.

24 O'Gieblyn, *God, Human, Animal, Machine*, 53.

25 Ibid., 72. O'Gieblyn claims the lecture was in 1957, but that's just when it was published; Huxley gave the lecture in 1951. See Christian Byk, "Transhumanism: From Julian Huxley to UNESCO: What Objective for International Action?," *Jahr* 12, no. 1 (2021): 141–162, https://hrcak.srce.hr/ojs/index.php/jahr/article/view/19376.

26 Here, Thiel seems to be talking about replicators, not transporters.

27 Maureen Dowd, "Confirm or Deny: Peter Thiel," *New York Times*, January 11, 2017, www.nytimes.com/2017/01/11/fashion/peter-thiel-confirm-or-deny.html.

28 Peter Thiel ("Changing the World" panel, Singularity Summit 2009, November 2, 2009), Vimeo, https://vimeo.com/7396024.

29 Arthur C. Clarke, *Rendezvous with Rama* (New York: Ballantine, 1973), 41.

30 Arthur C. Clarke, *The Exploration of Space* (New York: Harper and Brothers, 1951), 186.

31 James Cameron and George Lucas, "George Lucas on Star Wars Being Anti-Authoritarian | James Cameron's Story of Science Fiction," video, AMC+, YouTube, April 25, 2018, www.youtube.com/watch?v=fv9Jq_mCJEo.

32 Mark Caro, "'Star Wars' Inadvertently Hits Too Close to U.S.'s Role," *Chicago Tribune*, May 18, 2005, updated August 22, 2019, www.chicagotribune.com/2005/05/18/star-wars-inadvertently-hits-too-close-to-uss-role/.

33 J. W. Rinzler, *The Making of Star Wars: Return of the Jedi* (New York: Random House Worlds, 2013), 180. See also: Robert Vaux, Ben Hardwick, and Kieran Loughlin, "How a Real-Life War Inspired George Lucas' Star Wars Vision," *CBR*, last modified July 17, 2024, www.cbr.com/george-lucas-vietnam-war-star-wars-inspiration/.

34 Ursula K. Le Guin, "Author's Note" (1976), introduction to *The Left Hand of Darkness* (New York: Ace Books, 1969).

35 Charles Stross, "We're Sorry We Created the Torment Nexus," *Charlie's Diary* (blog), November 10, 2023, www.antipope.org/charlie/blog-static/2023/11/dont-create-the-torment-nexus.html.

36 Charles Stross, "Tech Billionaires Need to Stop Trying to Make the Science Fiction They Grew Up on Real," *Scientific American*, December 20, 2023, www.scientificamerican.com/article/tech-billionaires-need-to-stop-trying-to-make-the-science-fiction-they-grew-up-on-real/.

37 Stross, "We're Sorry."

38 Konstantin Tsiolkovsky, "Panpsychism, or Everything Feels" (1925), in Groys, *Russian Cosmism*; Konstantin Tsiolkovsky, "The Future of Earth and Mankind" (1928), in Groys, *Russian Cosmism*; Scharmen, *Space Forces*, 37; Young, *Russian Cosmists*, 151–152.

39 Stross, "We're Sorry."

40 Elon Musk (@elonmusk), Twitter, June 2, 2021, archived June 2, 2021, at the Wayback Machine, https://web.archive.org/web/20210602083709/https://twitter.com/elonmusk/status/1400008467822399490.

41 Stross, "We're Sorry."

42 Thiel Fellowship, accessed June 12, 2024, https://thielfellowship.org/.

43 Charles Duhigg, "Did Uber Steal Google's Intellectual Property?," *New Yorker*, October 15, 2018, www.newyorker.com/magazine/2018/10/22/did-uber-steal-googles-intellectual-property.

44 Reed Albergotti, "Ex-Uber Executive Anthony Levandowski Pleads Guilty to Trade-Secret Theft," *Washington Post*, March 19, 2020, www.washingtonpost.com/technology/2020/03/19/levandowski-guilty/.

45 Duhigg, "Did Uber Steal."

46 Adrian Daub, *What Tech Calls Thinking* (New York: FSG Originals, 2020), 3.

47 Blue Origin, "Blue Origin's Mission," February 1, 2019, video, YouTube, www.youtube.com/watch?v=1YOL89kY8Og.

48 Fridman, "Bezos."

49 Crawford, *Atlas of AI*, 235. See also *Encyclopaedia Britannica Online*, s.v. "V-2 rocket," accessed September 7, 2024, www.britannica.com/technology/V-2-rocket.

50 I didn't invent this term. Its origins are somewhat murky, but with its current definition it goes back at least as far as 2002. "Hactar," "Origin of the Term Engineer's Disease?," MetaFilter, June 29, 2016, https://ask.metafilter.com/297591/Origin-of-the-term-Engineers-Disease. See also this nice definition from Hacker News commenter "milesf" in 2015: "Technical people suffer from what I call 'Engineer's Disease.' We think because we're an expert in one area, we're automatically an expert in other areas." ("milesf," comment, Hacker News forum, December 30, 2015, https://news.ycombinator.com/item?id=10812804.)

51 Macicj Ccgłowski, "Moral Economy of Tech" (panel remarks, SASE, Berkeley, CA, June 26, 2016), https://idlewords.com/talks/sase_panel.htm.

52 Elon Musk, "Is Life a Video Game?," recorded at Code Conference 2016, video, Recode, YouTube, June 2, 2016, www.youtube.com/watch?v=2KK_kzrJPS8; See also Jason Koebler, "Elon Musk Says There's a 'One in Billions' Chance Reality Is Not a Simulation," *Motherboard*, June 2, 2016, www.vice.com/en/article/8q854v/elon-musk-simulated-universe-hypothesis.

53 All of these quotes are from Jorge Luis Borges, *Collected Fictions*, trans. Andrew Hurley (New York: Penguin Books, 1999).

54 Yuan (@QiaochuYuan), "sometimes it's nice to put on a movie . . ."

55 Cegłowski, "Moral Economy."

56 Friend, "Sam Altman's Manifest Destiny."

57 "OpenAI Charter," OpenAI, accessed September 6, 2024, https://openai.com/charter/.

58 An insular, self-reinforcing set of institutions and publication practices is a big part of this too, especially on the academic side. See Shazeda Ahmed et al., "Field-Building and the Epistemic Culture of AI Safety," *First Monday* 29, no. 4 (April 1, 2024), https://dx.doi.org/10.5210/fm.v29i4.13626.

59 Paul Krugman, "The Rich Are Crazier Than You and Me," *New York Times*, July 6, 2023, www.nytimes.com/2023/07/06/opinion/robert-kennedy-jr-silicon-valley.html.

60 Adam Ozimek (@ModeledBehavior), Twitter (now X), June 21, 2023, https://twitter.com/ModeledBehavior/status/1671485568360800259; Adam Ozimek (@ModeledBehavior), Twitter (now X), June 21, 2023, https://twitter.com/ModeledBehavior/status/1671486129747492864; Adam Ozimek (@ModeledBehavior), Twitter (now X), June 21, 2023, https://twitter.com/ModeledBehavior/status/1671486837959868417.

61 Friend, "Sam Altman's Manifest Destiny."

62 Paul Krugman, "The Paranoid Style in American Plutocrats," *New York Times*, August 28, 2023, www.nytimes.com/2023/08/28/opinion/columnists/covid-climate-cryptocurrency-plutocrats.html.

63 Anil Dash, "'VC Qanon' and the Radicalization of the Tech Tycoons," AnilDash.com,

July 7, 2023, www.anildash.com/2023/07/07/vc-qanon/. See also Homer Simpson: "If he's so smart, how come he's dead?"

64 Anil Dash, "The Tech Tycoon Martyrdom Charade," AnilDash.com, February 27, 2023, www.anildash.com/2023/02/27/tycoon-martyrdom-charade/.

65 See De Witt Douglas Kilgore, "'The Best Is Yet to Come'; or, Saving the Future: *Star Trek: Deep Space Nine* as Reform Astrofuturism," in *Black and Brown Planets: The Politics of Race in Science Fiction*, ed. Isiah Lavender (Jackson: University Press of Mississippi, 2014), https://doi.org/10.14325/mississippi/9781628461237.003.0003.

66 "Nichelle Nichols Remembers Dr. Martin Luther King, Jr.," StarTrek.com, January 16, 2023, www.startrek.com/news/nichelle-nichols-remembers-dr-king.

67 Benjamin Russell, "Far Beyond the Stars," *Incredible Tales of Scientific Wonder*, September 1953.

68 Mike Wall, "Earth from Space: 'Overview Effect' Could Help Troubled Country, Astrophysicist Neil deGrasse Tyson Says," Space.com, June 3, 2020, www.space.com/overview-effect-amid-us-crisis-neil-degrasse-tyson.html.

69 Stewart Brand, "'Whole Earth' Origin . . .," Long Now Foundation, accessed June 12, 2024, https://sb.longnow.org/SB_homepage/WholeEarth_buton.html.

70 Frank, *Light of the Stars*, 67–70.

71 M. H. Hecht et al., "Detection of Perchlorate and the Soluble Chemistry of Martian Soil at the Phoenix Lander Site," *Science* 325, no. 5936 (2009): 64–67, www.science.org/doi/10.1126/science.1172466.

72 Carl Sagan, *Pale Blue Dot: A Vision of the Human Future in Space* (New York: Ballantine, 1997), 6–7.

73 Stirone, "Mars Is a Hellhole"; Lex Fridman, "Elon Musk: Neuralink, AI, Autopilot, and the Pale Blue Dot | Lex Fridman Podcast #49," November 12, 2019, *in Lex Fridman Podcast*, YouTube, www.youtube.com/watch?v=smK9dgdTl40, transcript, https://lexfridman.com/wordpress/wp-content/uploads/2019/11/elon_musk_lex_fridman_2_transcript.pdf.

74 George Orwell, "Can Socialists Be Happy?," in *All Art Is Propaganda: Critical Essays*, ed. George Packer (Orlando, FL: Harcourt, 2008), 209.

75 Thiel, "Education of a Libertarian."

76 Adam Becker, "Junk Science or the Real Thing? 'Inference' Publishes Both," *Undark*, January 28, 2019, https://undark.org/2019/01/28/junk-science-or-real-thing-inference/.

77 Thiel, "Education of a Libertarian."

78 Kim Stanley Robinson, email to author, May 3, 2024.

79 Heidi Hackford, "The Valley and the 'Swamp': Big Government in the History of Silicon Valley," Computer History Museum, October 10, 2019, https://computerhistory.org/blog/the-valley-and-the-swamp-big-government-in-the-history-of-silicon-valley/; Henry Farrell, "Silicon Valley Paints Itself as a Hotbed of Free Enterprise. Here's

How the Government Helped Build It," *Washington Post*, August 17, 2019, www.washingtonpost.com/politics/2019/07/17/silicon-valley-portrays-itself-hotbed-free-market-enterprise-new-book-explains-how-government-helped-build-it/; see also Margaret O'Mara, *The Code: Silicon Valley and the Remaking of America* (New York: Penguin, 2020).

80 Raymond Lonergan, "A Steadfast Friend of Labor," in *Mr. Justice Brandeis: Great American*, ed. Irving Dilliard (Saint Louis, MO: Modern View Press, 1941), 42.

81 Ursula K. Le Guin, "Distinguished Contribution to American Letters Award Acceptance," November 19, 2014, American Rhetoric, accessed June 12, 2024, www.americanrhetoric.com/speeches/ursulakleguinnationalbookawardspeech.htm.

82 Ibid.

INDEX

Credit: Nick James

Adam Becker is a science journalist with a PhD in astrophysics. He has written for the *New York Times*, the BBC, NPR, *Scientific American*, *New Scientist*, *Quanta*, and many other publications. His first book, *What Is Real?*, was a *New York Times Book Review* Editor's Choice and was long-listed for the PEN Literary Science Writing Award. He has been a science journalism fellow at the Santa Fe Institute and a science communicator in residence at the Simons Institute for the Theory of Computing. He lives in California.